GW00713154

The Guinness Book of Why ❓

Donald McFarlan
and Ian Bishop
with additional material by
Robert Jackson

GUINNESS PUBLISHING

The authors would like to thank Robert Jackson for his contribution of the 'Why?'s' on pages 18, 32, 47, 62, 78, 85, 93, 104, 134 and 143, for this book.

First published 1994

© **Donald McFarlan and Ian Bishop 1994**

Published in Great Britain by Guinness Publishing Ltd., 33 London Road, Enfield, Middlesex.

Design and layout: John Mitchell.
Illustrations: Peter Harper, John Mitchell, Sarah Silvé and Matthew Hillier.

Printed and bound in Great Britain by Cox & Wyman.

'Guinness' is a registered trademark of Guinness Publishing Ltd.

A catalogue record for this book is available from the British Library.
ISBN 0-85112-574-3

Preface

We have enjoyed writing this book, although sometimes we would have liked to have had the chance to discuss it with you, the reader. We know what fascinates us but it is sometimes very difficult to decide if it is interesting to you. If there are ever any future editions of this book it would be extremely helpful to have some idea of the types of question which genuinely fascinated people outside the lonely little loft in Oxfordshire where this book was discussed, agonised over and eventually written. We cannot undertake to answer individual questions by post but if you have ideas for future 'why?'s, we would very much like to hear from you. We would find it interesting and it would help us a great deal. Please write to –

> Donald McFarlan and Ian Bishop
> c/o Guinness Publishing Ltd
> 33 London Road
> Enfield
> Middlesex
> EN2 6DJ

We will, of course, acknowledge the source of these ideas in future editions.

We hope you enjoy the book and we look forward to being in touch with you.

> Donald McFarlan and Ian Bishop
> Oxfordshire 1994
> Lat 51 Deg 51' 30" N
> Long 1 Deg 7' 10" W

To the Reader

More and more one comes to see that it is the everyday things which are interesting, important and intellectually difficult.

J. E. Gordon, *The New Science of Strong Materials*
(second edition, 1976)

This book is about a way of looking at the world and how it all fits together.

Everything that is observable – from the length of a queue in a supermarket to the shape of a honeycomb - is the result of a human decision or part of the vast web of cause and effect. Often it is a combination of both. (The managers of supermarkets have very well-researched and detailed information as to just how much patience the average customer is likely to have, while the bee's honeycomb is a beautiful piece of structural efficiency evolved over many thousands of generations.) Even the fact that the sentence that you are reading right now is vertically aligned on the left-hand side of the page but ragged on the right is not an accident. It is a conscious preference on our part as most research indicates that aligned (or 'justified' to use the printers' term) type is a barrier to ease of reading and understanding.

We would like to believe that if you get into a frame of mind of never accepting that something 'just is', the world becomes a much more interesting place. The trouble is that, rather like the construction of an onion, every 'why?' opens up another and another and so on, doubtless leading to the fundamentals of particle physics and some very abstract mathematics. That level of investigation is beyond the scope of this book and certainly beyond the abilities of the present authors. However, · the important thing is never to give up or admit defeat but to keep a curious mind alive. That is why the last entry in this book - 'Why are there no green mammals (or black plants)?' - is included. We don't know 'the answers' and we don't believe that anyone else would claim to, but that doesn't stop us thinking about such things. It shouldn't stop *you* either.

We would like to end this introduction by quoting from a 1981 television interview with amateur cracksman, part-time bongo drummer and Nobel Prize winning physicist, the late Richard Feynman, who perhaps more than most people embodied the spirit of 'Why?' (Even as a world famous professor of physics, just a few years before his death in 1988, he was still devising experiments about everyday things which puzzled him - for example, why a stick of dry spaghetti tends to break into three pieces rather than in half. Try it!) Here he is remembering his childhood and talking about his father who was a clothing salesman in New York City. (The italics are ours.)

'He taught me to notice things. One day I was playing with a little wagon for children to pull around. It had a ball in it, and I pulled the wagon and noticed something about the way the ball moved. So I went to my father and said, "Say Pop I noticed something. When I pull the wagon the ball rolls to the back, and when I suddenly stop, it rolls to the front. Why is that?" "That", he said, "nobody knows. The general principle is that things that are moving try to keep moving and things that are standing still tend to stand still unless you push on them hard. This tendency is called 'inertia', but nobody knows why it's true."

'*Now, that's a deep understanding. He didn't just give me a name. He knew the difference between knowing the name of something and knowing something, which I also learned very early.* He went on to say, "If you look close, you'll find the ball does not rush to the back of the wagon, it's the back of the wagon that you're pulling against the ball. The ball stands still, or as a matter of fact starts to move forward really, from the friction." So I ran back to the little wagon, set the ball up again and pulled the wagon from under it. Looking sideways I saw he was right - the ball never moved backwards in the wagon when I pulled it forward. It moved forward relative to the wagon, but relative to the sidewalk it moved forward a little bit and the wagon just caught up with it.

'So that's the way I . . . [learned things from] . . . my father, with no pressure, just lovely interesting examples and discussions.'

Contents

!

Why does a good lightning conductor almost never conduct lightning ❓

Lightning is caused by the negatively charged (that is with an excess of electrons) bases of thunder clouds violently releasing their charge to other clouds or to earth. Because like charges repel each other, the negative charge at the base of a thunder cloud induces a positive charge (in other words a lack of electrons) on the ground underneath it.

A modern lightning conductor consists of several sharply pointed iron rods connected to a thick copper wire running down the outside wall of a building to a large earthing plate buried in the ground. During the build up of an electrical storm the passing thunder clouds induce a strong positive charge in the rods at the top of the conductor and consequently - again because like charges repel - an equal and opposite charge at its other end. This latter charge runs safely away to earth through the iron plate.

The rods on the top of the conductor act in two ways. Because they are positively charged, the negative charge of the thunder cloud is attracted to them and runs to earth. (See *Why does milk turn sour?*) However, and much more importantly, because a pointed object develops a highly concentrated density of charge at its tip, what is known as an 'electric wind' of positively charged particles starts to develop. (This phenomenon can be demonstrated by the trick of blowing out a candle by moving a charged needle close to the side of the flame.) The electric updraft above the iron rods creates an umbrella of positive charge which keeps the

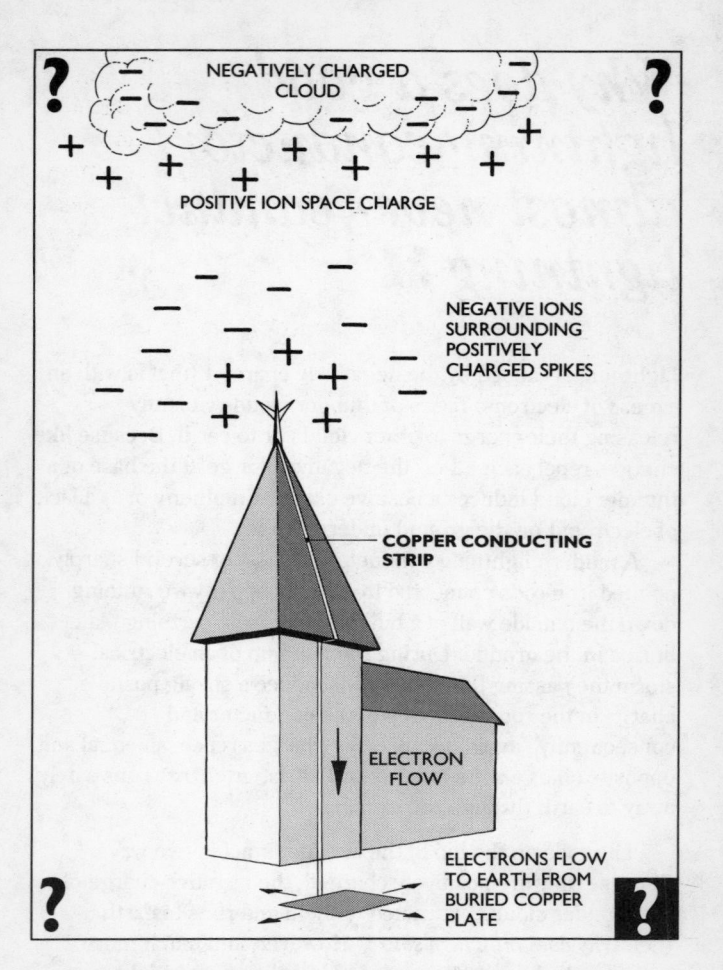

building perfectly safe without lightning ever going anywhere near the conductor.

Why is it called a lightning conductor, then?

If, however, when push comes to shove, there is a very sudden 'bolt from the blue', the conductor will do what its name suggests and quickly carry the charge safely to earth.

The reason that lightning can be so destructive is that a discharge can easily be in excess of 2000 amps and can generate a temperature several times that of the Sun's surface. Apart from the dangers of a direct hit to the mains electricity supply or a telephone line, the temperatures involved cause any damp in masonry to boil with explosive force, nails to melt and woodwork to catch fire.

One of the safest places to be in an electrical storm is actually inside a metal-bodied saloon car since charge will, as Faraday demonstrated, always remain on the outside surface of a hollow conductor. Although pure rubber is not a good conductor, there are sufficient impurities in car tyres (see *Why are car tyres black?*) for the lightning to run to earth. (Aircraft, which for a whole range of different reasons pick up charge when airborne, usually have especially conductive tyres so that any charge is quickly earthed without sparking on touch down). For the same reason the pilot or winchman of an air-sea rescue helicopter will always 'dunk' the lifting cable in the sea before attempting to hoist someone aboard. It might be regarded as something of a mixed blessing to be knocked out - or worse - by an electric shock just when you thought help was at hand.

!

Why would a wise yachtsman still prefer to use an anchor chain even though a modern rope is probably stronger ?

Probably because he wants a good night's sleep.

If a rope is used to attach the bow of the yacht to the anchor, it will tend to stretch in an almost straight line. In a rough sea this means that the bow of the boat will crash up and down on the waves as the rope rapidly alternates between slack and extreme tension. Not only is this very uncomfortable, but also the shocks transmitted along the rope can cause the yacht to drag its anchor.

If a chain is used, the weight of the links causes it to form a gentle curve (or catenary) between the two fixed points. As the bow of the boat moves up and down with the waves it has to act against the weight of the links which serves to dampen the violence of the motion. In addition, the anchor end of the chain will lie almost parallel with the sea-bed which, in a simple plough anchor, helps the anchor to dig in securely.

This 'catenary' effect (from Latin 'catena' - a chain) has given many a sailor a surprisingly comfortable night, even in the roughest of seas.

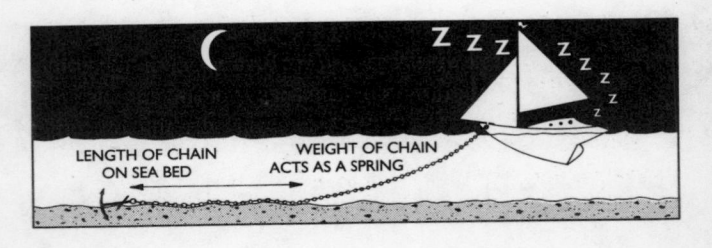

LENGTH OF CHAIN ON SEA BED

WEIGHT OF CHAIN ACTS AS A SPRING

Why is the red 'danger zone' on a car's water temperature gauge normally around 120° Celsius rather than at 'boiling point' ?

Pressure cookers are a demonstration of the fact that an increase in pressure raises the boiling point of a liquid. (See *Why can it take up to fifteen minutes to cook a 'three-minute' egg in villages high in the Himalayas?*) Almost all modern cooling systems in vehicles operate above atmospheric pressure. This means that the coolant can function quite happily at temperatures up to about 120 degrees Celsius and continue to cool the engine without boiling.

It is perfectly possible to cool an engine with a non-pressurized system (as was the case in all cars until the 1930s), but the trade-off is that more coolant has to be carried or a larger radiator must be used, in order to ensure that the temperature never approaches the 'atmospheric' boiling point of 100 degrees Celsius. As excess weight or a large frontal area are both undesirable, it is more efficient to pressurize the cooling system. The other advantage of running the radiator as hot as possible is that the greater the difference between the temperature of the radiator and that of the air, the quicker the radiator will lose heat.

However, there is an additional factor. The substance we normally refer to as 'antifreeze' is in fact a fairly sophisticated cocktail of chemicals with several properties. One of the common constituents of modern antifreeze is ethanediol (commonly referred to as ethylene glycol or simply glycol). When added to water, this has the effect not only of lowering the freezing point of the mixture but also raising the boiling point. Consequently, one can run even an unpressurized system to over 100 degrees Celsius at sea level.

Antifreeze or antiboil?

It is easy to think that because of the popular term 'antifreeze' it is for use in the winter months only, but it would be just as logical to call it 'antiboil'. There are other good reasons for using it all year round, as a corrosion inhibitor is also usually included in the mixture. This is especially important as a modern engine may have a whole range of dissimilar metals–cast iron, aluminium, brass, copper – which can lead to electrolytic reactions, within its cooling system.

Danger!

Why, then, is it so dangerous to open the radiator cap of a pressurized system when the engine has only recently been switched off? The coolant may well be at a temperature above the 'atmospheric' boiling point. When you open the cap the pressure is released and the coolant will instantly change from the liquid to the vapour state which can inflict a severe steam burn on the hand. A radiator cap should only be removed when you are certain that the coolant temperature has dropped to well below 100 degrees Celsius and even then it is worth putting a thick rag over the cap as a precaution.

!

Why does a heavily laden barge not increase the load on an aqueduct when it moves on to it ❓

For any object to float it must displace its own weight of water. Thus, when the barge moves onto the aqueduct an equivalent weight of water is moved off, and the total load on the structure remains the same. There might be a momentary increase in load as the water redistributes itself along the canal, but since (at low speeds) water is fairly fluid stuff, this is not significant.

Why do women (and some men) wear make-up ?

The obvious answer is that they wish to appear attractive to the opposite sex, but there is a far deeper and much more scientific reason than that.

It is never a bad idea to begin at the beginning; and the beginning, in this case, was when our ancestors descended from the trees. As they acquired an upright posture, their body underwent a profound structural change. The pelvic girdle was modified to support the upright gait, the foot lost its prehensile nature and became a rigid support organ, the folds of the peritoneum attaching the intestines to the abdominal wall adjusted to support the weight of the viscera, and the muscles of the buttocks became enlarged. This evolution into a creature that walked upright also brought about profound sexual changes; mating was achieved front to front, not back to front as in four-legged mammals. As a result, the mode of sexual selection was altered, the man's attention being focused upon his mate's breast and face. The man-woman relationship now involved the face, eyes and mouth of each partner.

No-one can say when the practice of adorning the body with trinkets and pigments to enhance natural beauty started, but there is evidence that Neanderthal people wore collars and bracelets made of shells and painted their bodies with ochre.

What began in prehistoric times as symbolic decoration was transformed into a high art form by the ancient Egyptians. But in their case the compounds that would today be called make-up were used mainly for medical purposes, rather than beauty-enhancing cosmetics. As lubricants and sunscreens, they used poultices and ointments made from the fat of sheep and oxen and from the oils of almond, sesame, castor and olive. The whole purpose, as one ancient papyrus states, was 'to make the joints of man more supple'. It seems that the Egyptians took such matters very seriously: during the reign of Ramses III (c 1198 –1167 BC) labourers working on the site of a necropolis at Thebes organised what we might now call a strike because their supply of ointment ran out. At an earlier date – about 1314 BC - the army of Seti I, campaigning in Syria, threatened rebellion unless their ointment ration was increased.

Other cosmetics also had important medicinal uses. Eye paint, for example, was first used in ancient Egypt as an insect repellant and as a safeguard against eye disease. Both men and women used kohl – lead sulphide – to darken their upper eyelids, while their lower lids were painted green with malachite (copper carbonate). The latter was also applied as a treatment for trachoma, a highly contagious disease of the eyelids. For 5000 years, before the advent of antibiotics, a solution of copper sulphate or carbonate was the only treatment for this ailment (and must have been pretty toxic).

From Egypt, these medicinal preparations spread to Greece, where they were called kosmetikos. Derived from a word meaning order or arrangement, it soon took on a new significance: to increase and maintain the body's health and beauty.

Kosmetikos became fashionable in ancient Rome, where the emphasis was on beauty preparations rather than medicines. Pliny the Elder (23 – 79 AD) described some of them: a soap that was used to dye the hair red, white lead that blanched the face, a liniment of almond oil mixed with milk, and a tooth powder made from powdered horn and pumice stone. Pliny was not in favour of such trappings, grumbling that their importation from Egypt and other Middle Eastern countries was draining the Roman exchequer. Despite this, he himself used a pomade of linseed oil and fat from the foot of an ox to combat the onset of wrinkles.

During the Middle Ages, the Christian Church was firmly opposed to physical adornment, but cosmetics continued in widespread use as medicines. In the 11th century, a Sicilian woman called Trottola di Ruggiero – mentioned by Chaucer in his prologue to The Tale of the Wife of Bath, and famous as a pioneer gynaecologist – prescribed cosmetic preparations as remedies for superfluous facial hair, body odour, sunburn, scalp diseases, acne and halitosis.

By such historical routes did cosmetology become entrenched in the modern world, its progress carried along by the needs of medicine as much as the desire to enhance personal beauty.

!

Why is the sky blue ❓

Colour is in many ways a psychological phenomenon and its effects are often deceptive. In fact the colour of the sky can vary from a deep indigo to orange or even red at sunset, but we tend to think of the 'natural' colour of the sky as being blue.

Since the Earth's source of light, the Sun, emits white light, it seems reasonable to wonder why the sky usually appears blue. White light is actually a combination of all the seven colours of the spectrum (red, orange, yellow, green, blue, indigo and violet, as seen in the rainbow) which are produced by light of different wavelengths. The attribute we call the colour of an object is due to a selective action on the light of the Sun, with different wavelengths of the combination which make up the white light being absorbed, reflected or scattered in different proportions. (The exceptions would be a perfectly transparent substance in which all of the light would be transmitted or a fluorescent substance – see *Why does tonic water have a bluish tinge?* and, for effects introduced by the sensitivity of the human eye – see also *Why does grass look white by moonlight?*)

An opaque, red object, for example, illuminated by white light absorbs all parts of the spectrum except red, which it reflects. A purple object reflects some red and some blue. A matt black object absorbs effectively all parts of the spectrum, while a white object reflects all.

When a beam of light passes through anything other than a total vacuum, there is going to be a certain amount of scattering. When light passes through the atmosphere from the Sun, it is scattered in all directions by the particles in the atmosphere. The short wavelength end of the spectrum (blue) is scattered to a very much greater extent than the longer wavelengths (red). On a clear day, when there is not much by way of dust or water droplets in the atmosphere and the scattering of the total amount of light is fairly limited, you will therefore see the sky as being blue. However, at sunset, when the light has to pass through a greater thickness of atmosphere as the Sun sinks towards the horizon (and when there is more dust in the atmosphere, especially at harvest time) the selective scattering will remove almost all of the blue short wavelength light, leaving only the yellow and red end of the spectrum visible at the Earth's surface.

If the Earth, like the Moon, had no atmosphere, the sky would always appear to be black, whether in day or night.

!

Why should we throw salt in the bilges of a new wooden boat **?**

Most of the water that finds its way into the bilges of a boat that is in good condition is either rainwater or condensation, rather than salt-water seepage. This provides a perfect environment for slime moulds to take over and attack the wood.

In fact, a saline solution is a near perfect disinfectant against this form of attack.

The next time you see a dinghy sunk at low - water mark during the winter, do not blame the fecklessness of the owner. It may be the best means of preserving it – especially when exposure to dry air might crack its timbers.

!

Why can it take up to fifteen minutes to cook a 'three-minute' egg high in the Himalayas ❓

At high altitudes atmospheric pressure is lower than at sea level. As pressure decreases, the boiling point of any liquid also drops and in these circumstances it would be perfectly possible to find water boiling at as low as 40 degrees Celsius. (A type of altimeter, known as a hypsometer, which consists of a beaker of water and a thermometer uses this fact to calculate altitude by measuring the temperature at which water boils - see also *Why should pilots be particularly careful if they fly from a region of high pressure to a region of low pressure?*) Since the process known as cooking involves transferring heat energy from the cooking medium (say, water) to the food, it obviously takes longer if the water is at a lower temperature. Boiling an egg at high altitudes is the equivalent of trying to cook one in tepid water at sea level. Such was the seriousness of the problem that, in the last century, tables were published giving the time needed to boil an egg at all of the major Indian hill stations in the Himalayan foothills.

We can see the same principle, but in reverse, in the domestic pressure cooker. In a pressure cooker the whole cooking environment is sealed-in. Because nothing can escape, the internal pressure rises under expansion and the boiling point consequently rises. In a typical modern pressure cooker the internal pressure can safely be made to rise to about twice that of the surrounding atmosphere. At this pressure the boiling point of water is around 120 degrees Celsius, which greatly reduces the cooking time. (A safety valve is, of course,

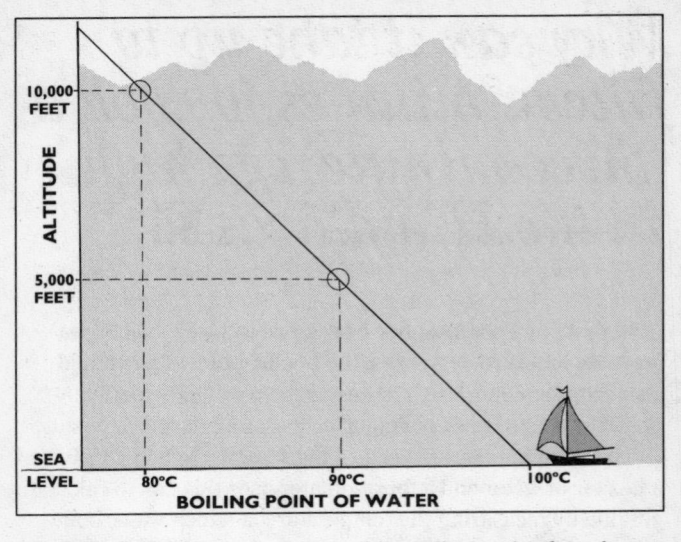

fitted to allow excess pressure to 'blow-off steam', otherwise the cooker would eventually explode!)

The relationship between temperature and cooking time is not a linear one (ie a 10% increase in temperature does not give a 10% saving in time), and a pressure cooker can reduce cooking time by over half. Apart from the time-saving benefit, there is also the advantage that vegetables cooked in this way retain more flavour, texture and nutritional value than if they were boiled for longer at atmospheric pressure.

Cup of tea?

Tea made at high altitude with water which is boiling at less than 100 degrees Celsius has a different taste from that normally experienced at sea level.and may not be to the liking of all. Indeed the explorer and mountaineer, Raymond greene, (brother of the novelist, Graham) described tea infused at Himalayan altitudes as being in 'a miserable state of lukewarmness'. You can try this for yourself by infusing some tea in water at around 40 degrees Celsius.

Why are car tyres black ▉

Because car tyres have been almost universally coloured black for the majority of this century it is easy to fall into the assumption that this is their 'natural' colour. In fact the colour of rubber after vulcanization (a process which hardens and toughens the natural rubber by mixing it with sulphur) is a yellowish grey. If you look at photographs of very early motor cars, you will see that their tyres are much lighter in colour than we would nowadays expect.

The rubber compound in today's tyres is a very sophisticated piece of chemistry indeed with sometimes over twenty different ingredients in a subtle balance contributing to that particular mix.

However, for many years now, one of the most important ingredients has been carbon black. (The soot produced by a badly trimmed candle or oil-lamp is a form of carbon black, hence its old-fashioned name of 'lamp black'.) The type and proportion of carbon black added to the mix allow the tyre designer to tune the performance characteristics of the finished tyre.

In general, an everyday road-going tyre has a fairly low level of coarse-grained carbon black added to it, while a high performance racing tyre will have a relatively high level of fine-grained carbon black.

Because the finer grains have a greater total surface area, they generate more heat through internal friction and the tyre will run much hotter. This suits the required characteristics of a racing tyre where maximum traction is the primary objective with the tyres becoming almost 'gummy' once they have reached racing temperature. (The reason you will see racing drivers making quite violent swerves on a warm-up lap is to raise the temperature of their tyres, which obtain their maximum grip at around 90 degrees Celsius. Fresh tyres are kept in electrically-heated warmers before being fitted during a race.)

On a road tyre, however, considerations such as low rolling-resistance (for fuel economy) and a low rate of wear are much more important and a colder-running compound will be used. As ever, of course, it is a question of balancing many different requirements and a tyre appropriate for a low-powered delivery van will use a very different compound from that intended for a sports car.

!

Why should you avoid putting unrotted compost on your flowerbeds ?

There are several reasons for digging compost – which is nothing more nor less than vegetable material in various stages of decomposition - into the soil. One of these is that the nitrogen compounds in the compost are decomposed by bacteria in a complex series of reactions into ammonia and the nitrates which are essential nutrients for plants. However, at certain stages of these reactions the bacteria can actually take out nitrates from the soil and thereby negate the object of the exercise. A successful compost should be very well rotted – generally for about a year – on a compost heap before it goes anywhere near your garden plants.

Why do lemmings have such an undeserved reputation for voluntary mass suicide ?

One of the the more popular journalistic metaphors for irrational mass human behaviour is to describe it as lemming-like. This is misleading and probably unfair to the lemming. The lemming (*Lemmus lemmus* or *Lemmus norvegicus*) is a small rodent which lives in the sparsely-vegetated uplands of Norway and Sweden. (Although related species do exist in other similar habitats in the Northern Hemisphere the lemming behaviour of popular imagination is mainly Scandinavian.) Their diet is entirely vegetarian and consists

mainly of grass roots, mosses and lichens. A lemming brood can consist of anything between three and eight young and there are *at least* two broods every year. Given the fluctuating nature of their breeding and the sparseness of their environment, which can be covered in snow for a large part of the year, it is hardly surprising that the size of the population and the limitations of the food-supply sometimes get out of kilter. When this happens they may descend in vast armies on the cultivated lowlands in the search for food, advancing slowly in a constant direction, overcoming all obstacles and often swimming lakes of several miles breadth. They never turn back and eventually some of them will inevitably reach the sea. The instinct which drives them on here is nothing to do with self-destruction, but is precisely the same one which led them to (successfully) cross lakes in earlier stages of the migration.

Because of their irregular and seemingly inexplicable sudden appearances at the coast, lemming behaviour has long given rise to different speculations. One such was that they fell from the clouds. Another was that they were seeking out their original home in the drowned city of Atlantis. Perhaps some modern beliefs are scarcely more rational.

!

Why does ice form in a carburettor even when the air temperature is several degrees above freezing point ❓

Carburettor icing is an extremely baffling phenomenon the first time it is encountered in a car. Typically it happens on a damp, misty day when the air temperature is a few degrees above freezing. (When the air temperature is actually below freezing, the problem does not arise since any moisture in the air is deposited as frost on any available surface.)

The engine starts perfectly well but then progressively loses power and eventually 'dies'. Attempts to restart are initially unsuccessful. However, if the engine is left alone for a few minutes, the ice in the carburettor will melt and the engine will restart with a splutter as the water clears from the carburettor. The whole cycle of events will then tend to repeat itself ...

While this is merely inconvenient in a car, it is obviously extremely dangerous in petrol-engined light aircraft which are particularly at risk by dint of having to operate at constantly changing levels of humidity, temperature and pressure. Carburettor icing is suspected of playing a role in many aircraft accidents, but, like the proverbial dagger made of ice in a murder mystery, the evidence has vanished by the time the inspector arrives.

What causes this phenomenon?

When a liquid vaporizes it draws heat from its surroundings. This can be observed by placing a drop of a volatile liquid such as aftershave on the back of your hand. As the aftershave evaporates, you can feel a cold area on your skin. (See *Why do we add ice cubes to our drinks?*)

Similarly, on a damp day, you can see a small ring of frosting around the nozzle of an aerosol scent spray immediately after it is used. Again, the contents of the can are changing from the liquid to the vapour state, but here the process is accelerated by the large drop in pressure between the inside of the can and the atmosphere.

Exactly the same thing is happening inside a carburettor venturi which depends on a pressure drop to draw the fuel into the carburettor and vaporize it in the fast-moving air through the venturi.

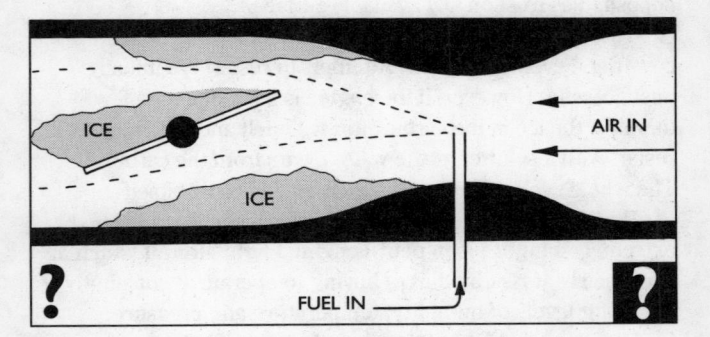

Any moisture in the air will tend to build up as ice around the 'waist' of the venturi. This progressively strangles the engine and in extreme cases interferes with the operation of the throttle butterfly valves.

In light aircraft, the pilot's remedy is to select an alternative air source which supplies hot air to the carburettor, usually from the vicinity of the exhaust manifold.

In the case of road cars, there are often alternative winter/summer settings on the air filter housing which duct heated air - usually from above the exhaust manifold - to the carburettor if correctly set up at the start of cold weather.

It might seem simpler always to draw warm air into the carburettor but (see *Why does a car seem to perform better on a cold, damp day?*) it is better to use the coldest induction charge of air that you can get away with.

Why do certain drinks such as Pernod and ouzo change from clear to cloudy when you add water to them

These drinks all contain certain naturally occurring essential oils (mainly of the terpene group) which are added for their flavour and aroma. These are soluble in alcohol but not in water. When you add water the alcohol concentration obviously drops and these oils are precipitated out to form an emulsion in the alcohol water-mix.

Why doesn't a spider get caught in it's own web ❓

To answer this question, we must first of all examine the miracle of engineering that constitutes a spider's web.

In terms of tensile strength, the silken strands that make up the web are far stronger than steel and second only to fused quartz; a strand can be stretched to a fifth of its length before it snaps. In fact, a strand that is visible to the naked eye is more in the nature of a cable, composed of several interwoven tiny threads; individual threads may be no more than a millionth of an inch thick. The material is very common, for a single acre of territory in the British Isles may be inhabited by up to two and a quarter million spiders.

Each spider has nipple-like spinnerets, usually three, located near the end of the abdomen. On each spinneret there are several tiny orifices through which the secretion that forms the silken threads is expelled from the spider's glands. As it spins its web, the spider brings the tips of the spinnerets together so that the streams of secretion unite to form a single thread.

Spiders that live out of doors spin a type of web known as the orb, a masterpiece of symmetry that takes on a beautiful aspect when spangled with early morning dew. It is the female spider who does all the work in constructing the orb. Using abdominal pressure, she forces silk to flow outwards from one of six different glands in her abdomen, initially attaching one end of the first strand - known as the bridge – to a stem of grass, leaf or twig. Dropping down with the strand, still weaving, she runs across the ground to another high spot, pulls the strand tight, and attaches it firmly in place with a spot of sticky glue from another of her glands.

Once the first – and always horizontal – strand is fixed, the spider drops two plumb lines of silk, one at either end, and constructs a second bridge lower down to form a framework. Within this framework she spins a series of radii with a hub at the centre. And this is the clever part: as she builds the radii, only the outer radius is coated with sticky globules of a substance produced by a specialist organ called the ampullate gland. Only when her task is done does she deposit more sticky globules at intervals on the other radii, leaving spaces to step between.

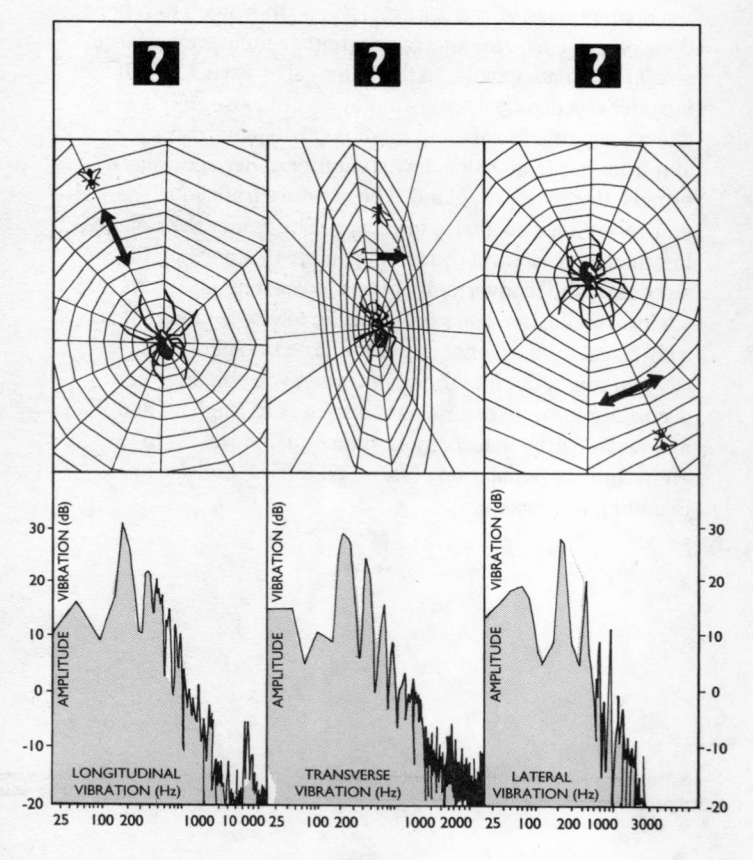

With the web built, the spider constructs a nest nearby, deftly rolling a leaf and making herself comfortable inside on a bed of silk. She may, after all, have a long wait before her first victim arrives. As a final touch, she spins a warning strand of silk between the nest and the radius closest to the hub of the web, so that when something lands on it the strand vibrates. She then hurries to the centre of the web along the untreated warning strand, stopping frequently to analyze just what it is she has to deal with.

Spiders have very poor eyesight, so they need to rely on their other senses to determine whether their prey is large, small or unmanageable. In the latter case - particularly if the intruder is a deadly enemy - it may be cut loose. If it is large, like a wasp or a hornet, but inedible, the spider can squirt silk at it from a safe distance, using another of her specialised organs, the aciniform gland. The creature trapped in the web enmeshes itself as it tries to escape. The spider then makes her final approach via the built-in stepping stones on the web strands; if she inadvertently puts a foot wrong and steps into a sticky globule, her body produces an oily secretion that acts as a chemical thinner, enabling her to free herself.

You might not like them, but when you consider that according to some estimates the spiders of England and Wales annually destroy insects more than equal in weight to the entire human population of those countries, there is no disputing their value.

!

Why will a pendulum clock tend to run slow in the Summer ❓

The duration of swing of a pendulum is a constant governed by the pendulum's length. In other words, any two pendulums of the same length (measured from the pivot at the top to the centre of gravity of the bob at the bottom) will behave in the same way irrespective of their weight or what they are made of. (This is assuming no variations in the force of gravity. There are, in fact, small but significant differences in gravitational force at different places on the Earth's surface - and the makers of early pendulum clocks were aware of them. In general, the pull of gravity decreases the closer you are to the Equator and a pendulum will swing more slowly.)

No matter what it weighs, a long pendulum has a slow swing and a short pendulum a rapid one.

In early eighteenth-century clocks regulated by a pendulum, the rod of the pendulum tended to be made of steel. While steel is an excellent material for this purpose in that (unlike wood) its behaviour is not affected by changes in humidity, it is strong and does not significantly stretch over the years, it does, however, expand with a rise in temperature. With a pendulum of 99.42cm length from the pivot to the centre of gravity of the bob (the theoretical length to produce a swing with a period of duration of swing of one second) an increase in temperature of 30 degrees Celsius will cause the clock to lose slightly over 12 seconds a day.

In an age before broadcast time signals or indeed any other means of communicating the 'real' time over long distances accurately, this created considerable inconvenience.

The first pendulum which compensated for temperature

was an extremely ingenious device invented by George Graham in 1722. This used a column of mercury enclosed within a glass tube in the bob. As the temperature rose and the pendulum lengthened, the mercury expanded and was forced to rise up the glass tube, raising the centre of gravity of the bob and thus keeping the effective length of the pendulum the same.

Another form of compensating pendulum was invented in 1726 by the great English clock maker, John harrison. This 'gridiron' pendulum consisted of rods of steel and brass in which the rods of brass expanded upwards from the bob raising the centre of gravity and compensating for the lengthening steel.

The need for such ingenuity largely came to an end, however, with the invention in the late nineteenth century by Charles Guillaume of invar, an alloy of nickel and steel. This had a coefficient of expansion of less than one tenth that of steel and allowed relatively simple brass compensating bobs to be used.

!

Why should you never put ripe bananas next to cut flowers **?**

Tips on helping cut flowers to remain fresh usually include such advice as keeping them away from radiators, stuffy rooms and direct sunlight - all of which seems fairly obvious. However, another piece of advice is sometimes given - to keep them away from the fruit bowl. Why?

Many species of fruit give off tiny quantities of ethene gas when they are ripe. This gas has the effect of accelerating the ripening of other fruit and vegetation. This is why people sometimes put a ripe tomato alongside a cluster of green tomatoes which they want to ripen quickly and is one of the reasons why one bad apple spoils the barrel.

Ethene gas has exactly the same effect on flowers and a market gardener will always be careful to keep flowers away from a shed of ripening apples. The effect at home is probably small, but could be significant in a poorly-ventilated room. Very ripe bananas are one of the more productive sources of ethene and leaving them next to a vase of cut flowers is probably best avoided. Incidentally, using a ripe banana will be a more efficient means of speeding up the ripening of your tomatoes than a ripe tomato.

!

Why do deep-sea divers breathe a mixture of helium and oxygen rather than natural air ?

The major constituents of air are nitrogen and oxygen in a proportion of roughly four to one by volume. Nitrogen is a fairly inert gas and in normal breathing passes into the body without causing any chemical change or other untoward effects.

However, water pressure increases proportionately with depth and to sustain normal breathing a diver has to be supplied with a breathing mixture at an equivalently increased pressure. At over 30 metres depth, for example, a diver will be breathing a mixture at approximately four times atmospheric pressure - and if this mixture is air, four times the quantity of nitrogen.

Nitrogen is absorbed by the body's lipids (fatty tissue) much more readily than by the rest of the body. As the brain and central nervous system have a lipid content of around sixty per cent, they rapidly become saturated with the gas, disrupting their normal function.

The resulting condition is nitrogen narcosis, also known as 'narks' or the 'raptures of the deep'. The condition is very similar to drunkenness, with reason and physical dexterity becoming increasingly impaired.

The other danger associated with breathing a nitrogen mixture under pressure is decompression sickness or 'the bends'. If the diver ascends too rapidly the nitrogen in the saturated lipids will expand under the reduced pressure and form bubbles in the brain, spinal cord and joints, causing a wide range of symptoms including convulsions, paralysis and intense pain.

These two conditions can be avoided if a diver breathes a mixture of helium and oxygen in place of air. Helium is inert so does not react with any of the body tissues and is much less readily absorbed by them than is nitrogen. However, it conducts heat much more readily than nitrogen which can lead to the body losing heat to the surrounding water very rapidly so a diver breathing a helium/oxygen mix would normally wear a heated suit. The other disadvantage is that speaking through such a mixture produces a 'Donald Duck' voice familiar as a party-trick to those who have been let loose with helium-filled balloons at social gatherings.

!

Why do we measure the strength of spirits in 'degrees proof' ❓

Proof spirit was originally defined in the UK as the weakest solution of alcohol (ethanol) in water which would burn with gunpowder and ignite it (if the spirit was too weak the alcohol would still burn off, but the excess of water would prevent the gunpowder igniting). It was standardized under George III and represents a concentration of ethanol of 49.28% by weight or 57.10% by volume.

This leads to a great deal of confusion and the confident assertion of some unbelievable rubbish in saloon bars world-wide. Tales of mega-potent exotic firewaters of '110% alcohol' are not uncommon. 'Undiscovered' tavernas on Greek islands are generally involved.

While the strength of beers and wines is usually given as the percentage of alcohol by volume (presumably because it gives a more impressive or more cautionary figure than by weight), spirits produced in the UK were until very recently conventionally measured in degrees proof. A spirit labelled 100 degrees is proof spirit - in other words 57.10% alcohol by volume. More commonly nowadays (our nineteenth-century ancestors had more robust palates), a typical whisky will be 70 degrees proof, or 70% of proof spirit – in other words 39.97% alcohol by volume. The legendary 110% concoctions are probably 10 degrees 'over proof' or 62.81% alcohol by volume – which does not sound nearly as impressive.

In the US, proof spirit is more simply defined as a solution of 50% alcohol by volume, which, along with the US gallon and pint, seems like yet another example of Americans giving short measure to British eyes.

In passing, it is always important to make it clear when we are talking about percentages by weight and percentages by volume. In the case of relative quantities of alcohol and water it does not make a huge difference. Sometimes it can, however. The composition of alloys, for example, is almost always given as percentages by weight. This might lead you to be surprised that cast iron with as little as 4% carbon in it is such a radically different material from the purer wrought iron. When you remember that a carbon atom is one fifth of the weight of an iron atom and that by volume the cast iron is twenty per cent carbon, it becomes less surprising.

Why should you never oil a traditional grandfather clock ?

One of the simplest ways of hardening the surface of a metal, as our Bronze Age ancestors would have confirmed, is to hit it with a hammer – 'work hardening'. This is because the repeated blows drive the many defects in the crystalline structure of the surface deeper into the metal. Indeed, a traditional clock maker would still work harden brass cogs before assembling the clock.

This process continues, in a small way, in the workings of the clock as cog strikes upon cog day by day. By oiling the mechanism, all you are doing is to damp out the minute blows and producing a wickedly abrasive paste as dust and grit accumulate in the oil over the years. This will obviously wear out the clock. If you refrain from using the oilcan, the clock should be good for several generations.

Why were pot plants once taken out of an invalid's bedroom at night ❓

Unlike animals, all green plants make their own food (glucose) in a process known as photosynthesis. In this series of reactions, carbon dioxide from the air and water from the plant's roots are combined to form glucose in the plant's cells while oxygen is given off through the plant's leaves. In this respect, plants behave in a reverse manner to us in that we breathe in oxygen and breathe out carbon dioxide. Necessary for the process in plants, however, is sunlight whose energy is used by the green pigment in the plant, chlorophyll, to fuel the reactions. When there is sunlight about, the plant and the invalid are not in any way competing for resources.

However, in the absence of sunlight this process stops and the plant may actually absorb tiny amounts of oxygen. It could be argued that, in a minuscule way, the invalid and the plant are in competition and hence the old practice of removing plants at dusk. However, in any twenty-four hour cycle a plant is a net producer of oxygen – otherwise it would die through glucose starvation – and on balance it is now felt that nurses have better things to do with their time than to cart vegetation around.

❗

Why does an engineer avoid using a pencil to mark metals ❓

Metals react electrolytically with each other and fall into a reactivity series (See *Why does the salting of road surfaces in winter make steel-bodied cars rust more quickly?*) As well as metals, certain non-metals react electrolytically and have their place in the reactivity series. Carbon is one of these and falls into the series below tin. Not only is it below tin but, while the metals in the series all lose electrons when they react so that the electrolytic action between them is merely relative, carbon actually gains electrons when it reacts with metals. This means that the electrolytic action is very powerful and indeed is the source of the electrical energy in some types of torch battery which have a carbon rod as the positive electrode and a zinc casing as the negative electrode. (If you remember this you will never make the mistake of putting the batteries in the wrong way round in a pocket radio even if the polarity markings have worn off the case of the batteries. Whatever the chemical

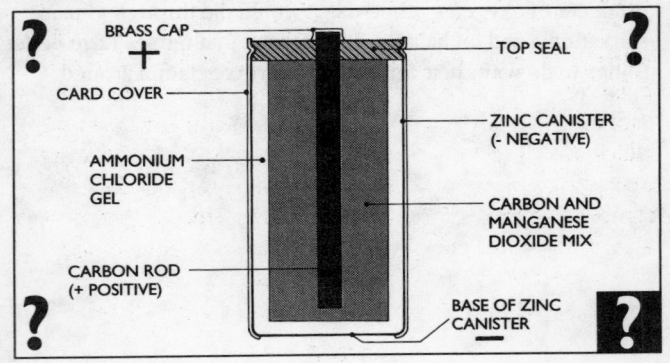

BRASS CAP
+
CARD COVER

AMMONIUM
CHLORIDE
GEL

CARBON ROD
(+ POSITIVE)

TOP SEAL

ZINC CANISTER
(- NEGATIVE)

CARBON AND
MANGANESE
DIOXIDE MIX

BASE OF ZINC
CANISTER

reaction used in the battery, convention now dictates that the raised 'blip' on one end of the battery will always be the positive terminal since it originally covered the end of the carbon rod in the early zinc-carbon batteries. A single zinc-carbon battery, or more correctly 'cell', produces a voltage of about 1.5v.)

The 'lead' in an ordinary pencil is made up of graphite (a form of carbon) and china clay (aluminium silicate). The harder the pencil, the higher the proportion of graphite. While it might be regarded as bad practice, it probably wouldn't matter too much to mark steel with a pencil as the steel is generally going to be protected by cleaning and then painting it later. However, when it comes to a highly reactive metal such as aluminium alloy sheet, which may be used in its unpainted state on an aircraft wing, this would be regarded as a fireable offence. The reason aluminium can be left unpainted is that it is protected by a thin layer of oxide which prevents further corrosion. (The statue of Eros, in Piccadilly Circus in London, which has lasted since 1893 is a classic example of the durability of aluminium provided the surface is not disturbed.) By marking it with a pencil you are scratching through this layer and leaving the aluminium directly in electrolytic contact with the graphite. In the fullness of time the graphite will etch its way through the aluminium – potentially disastrous if the shape you drew in the first place was a circle.

To avoid such problems, an engineer will always mark metals with a sharp-pointed scribing tool if marking a line which is later to be cut or will use a special marking dye which contains no carbon. The scriber merely scratches the surface of the metal, which in the case of aluminium quickly 'heals over' with a layer of protective oxide, while the dye will not set up any electrolytic reactions. In more critical work - for example on highly-stressed aircraft skins - the engineer will cover the alloy sheet with a protective film and draw his marks on that. The film is then removed after all drilling and cutting has been completed.

!

Why does a fish benefit from having a flabby skin ❓

If you were to make an exact replica of a real trout out of wood it would generate more drag as it passed through water than the living example. Why?

When a body passes through a fluid a resistance or drag is felt. This resistance is at its lowest when the flow of fluid over the body is smooth. As the body goes faster, areas of turbulence build up which greatly increase the resistance.

This is where the living trout has an advantage over the wooden model. As the turbulence builds up the trout's skin gives slightly at the point where the turbulence is worst, and absorbs its effects, creating a smooth flow of fluid over the fish's body.

Attempts to mimic this effect in the hulls of racing yachts by covering them in a flabby liquid-filled membrane have been outlawed by all of yachting's governing bodies. The performance of the resulting vessels has been so spectacularly improved that no conventional yacht would ever have won a race again.

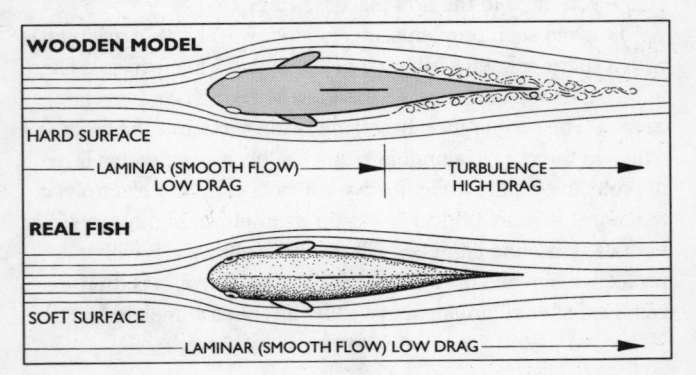

WOODEN MODEL

HARD SURFACE

LAMINAR (SMOOTH FLOW) LOW DRAG — TURBULENCE HIGH DRAG

REAL FISH

SOFT SURFACE

LAMINAR (SMOOTH FLOW) LOW DRAG

Why should pilots be particularly careful if they fly from a region of high pressure to a region of low pressure ?

While both Descartes and Pascal predicted that atmospheric pressure would decrease with an increase in altitude, it was not until Pascal persuaded his brother-in-law to carry a mercury barometer - in which a column of mercury is supported in a vertical tube by atmospheric pressure - to the summit of the Puy de Dôme (altitude over 3000 feet) in the Auvergne Mountains of central France, that this phenomenon was effectively measured.

Pascal's experiment showed that there is a drop of roughly 25 millimetres of the supported mercury for every 1000 feet ascended and hence that altitude can be measured through a difference in pressure. To this day, altimeters rely on this fact, although now an aneroid (from the Greek meaning 'without liquid') barometer would be used. This compares the difference between an enclosed vacuum and the external pressure.

But what's this got to do with pilots?

A pressure altimeter as fitted in an aeroplane is simply a very sensitive aneroid barometer calibrated to indicate height instead of atmospheric pressure. Herein lies a problem, however. Since atmospheric pressure varies as the weather changes, an altimeter has to be adjusted to take into account the local pressure at the time and place from which the aircraft wishes to take off, as well as the altitude of the airport. (This can cover a very wide range indeed. For example, the world's highest landing field, Lhasa in Tibet, is at an altitude of over 14 000 feet, while the lowest, El Lisan on the shores of the Dead Sea, is almost 1200 feet below sea level.)

As well as setting the altimeter for the point of departure, the pilot also, and certainly more crucially, needs to know the correct altimeter setting for the airport at which he wishes to land – where not only altitude but also atmospheric pressure may be radically different. Nowadays a radio call to the air traffic control at the destination airport will conveniently solve the problem. There, the controllers will check their own extremely accurate barometer and advise the pilot accordingly.

However, the considerable dangers of flying over mountainous territory where weather, in any case, is liable to be more variable, can be seen in the following, perfectly realistic, example.

Suppose you are flying with a constant altimeter setting of 4000 feet from an area where a barometer at sea level is reading a relatively high pressure of 1030 millibars to an area where a barometer at sea level is reading a relatively low pressure of 1000 millibars. (The millibar is one of the modern units for expressing pressure and is equivalent to about 0.75 millimetres of mercury). There is a 3500 foot mountain on your route in the area of low pressure. Given that one millibar is roughly equivalent to 30 feet of altitude, your altimeter – fooled by the drop in pressure into thinking that height is being maintained – will still be perfectly happily reading 4000 feet when you hit the mountainside 900 feet beneath the summit.

Hence the pilot's standard mnemonic, 'High to low (ie pressure), careful go'.

Why do some sea creatures glow ❓

Ever since Christopher Columbus was the first European to discover it during his westward voyage in 1492, the Sargasso Sea - the elliptical area of Atlantic Ocean southeast of Bermuda, with its islands of floating sargassum weed - the luminescent qualities of sea creatures have fascinated generations of seafarers.

In the Atlantic spring, a myriad creatures graze on the weed with insatiable hunger. Among them are krill, shrimp-like crustaceans no more than an inch and a half (4 cm) long. Themselves the food for almost every sea-dweller that can find them, from gulls to blue whales, they attack the weed with a rotary mouthpiece and strip it like locusts.

If they are disturbed, a curious thing happens: they burst into light. The glow produced by half a dozen krill in a jar is strong enough to read by. The reason is that krill have bioluminescent organs known as photophores on the underside of their bodies; they are glandular in origin and produce light by means of a chemical reaction.

Krill are only one of many species of sea creatures which are luminous. They share an ability that humans have always lacked: the means of converting chemical energy to radiant energy without generating heat. The process is virtually 100% efficient and, as heat is absent, the luminosity that it produces is called cold light.

Cold light is produced by some land creatures, such as fireflies, but it is the oceans that teem with it. The biochemical events that produce it are complex and are not common to all species, but the process involves two essential light-emitting components – an oxidizable organic molecule named luciferin and the enzyme lucifrase. When luciferin is oxidized by molecular oxygen, the reaction is catalized by the enzyme luciferase, with a resulting emission of light. This light emission continues until all the luciferin is oxidized.

Over the years, marine biologists and chemists have reached a clearer understanding of how bioluminescence occurs, but the reason why sea creatures turn on their light varies. In the majority of cases, it appears to be associated with the protection and survival of the species, either for sexual attraction, the luring of prey or self-defence.

For example, the photophores of many deep-sea fishes are arranged on the belly and lower sides, so that light is emitted downwards and outwards. The theory is that the light of the photophores matches the intensity of light filtering down from the surface, so that the fish's shadow is not readily visible to a predator approaching from beneath.

The deep-sea angler fish uses light as a lure. It cruises with jaws agape, its first dorsal spine turned forward to form an elongated rod with a luminous organ at its tip to attract its prey, which is then engulfed in the large maw.

Other uses of bioluminescence are not so easy to understand. One light-producing mollusc, a bivalve called the common piddock, bores holes in the rock around Europe's coastlines and emits a glow as it does so, although nobody knows why.

The piddock was once widespread and a table delicacy in the Mediterranean, although pollution has reduced it to a fraction of its former numbers. It uses its serrated shell to carve circular holes into hard rock, and its Latin name, Pholas dactylus, means 'finger in the hole'.

Scientists have discovered that the piddock's luminescence is the result of the reaction of a photoprotein, another protein and molecular oxygen. The glow it emits is dull and unearthly, but it can be stimulated to greater brightness if the creature is gently stroked. The piddock's equivalent of purring, perhaps?

Why does a hot air balloon rise (and why does a bicycle pump get hot) ❓

You might think that the lightest possible vessel would be a totally empty – or evacuated – one, and indeed you would be absolutely correct. The first in any way theoretically sound proposal for a flying machine followed this principle. It was proposed in 1670 by the Jesuit priest Francis Lana. He described an 'aeronautical engine' which would be supported by four evacuated copper globes each of about eight metres in diameter. Unfortunately, it was totally impracticable in that it could never, under any circumstances, be built.

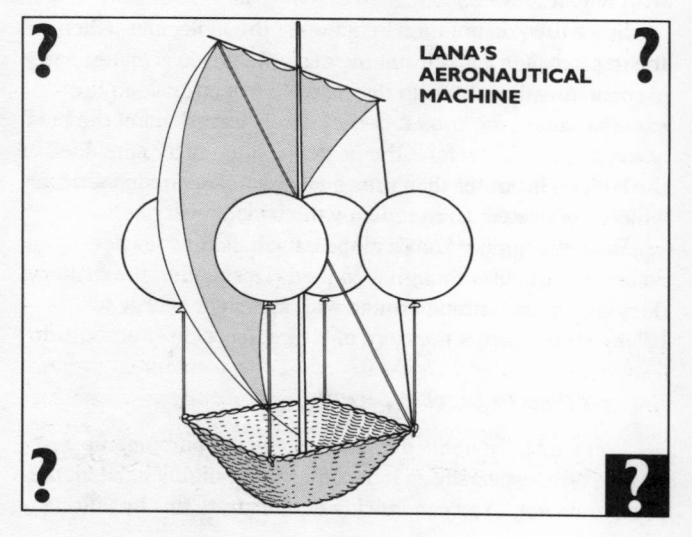

LANA'S AERONAUTICAL MACHINE

His calculations were correct and indeed if such a scheme were ever possible the globes would be lighter than the volume of air they displaced and consequently they would rise.

However, the flaw in his scheme is that since, for the sake of lightness, the copper skins would have to be much less than one millimetre thick (as he proposed), the globes would collapse under atmospheric pressure during evacuation, even if they were able to support their own weight during construction. Nevertheless, a profitable line of enquiry had been established - how to contain something which was less dense than the atmosphere without the container collapsing.

By definition, when a gas is heated the molecules within it acquire more energy and randomly rush around more violently in all directions taking up more and more space as the temperature increases. In other words, the gas expands. (This is formulated in Charles' law named after the physicist and pioneering balloonist, J A C Charles. Oddly enough, the first balloon he helped to construct was a hydrogen, rather than hot air, device.)

Since there is nothing 'in between' the molecules, when the gas expands a given volume of it contains less matter and is consequently lighter. In the case of a hot air balloon the excess matter - the cooler, denser air - is forced out of the hole at the bottom. Therefore, the heated volume of air contained in the balloon is lighter than the equal volume of atmospheric air which it displaces. Consequently the balloon will rise.

However, unlike Lana's globes, the balloon does not collapse because, although it contains less matter, the matter it does contain is bashing around with sufficient energy to balance the external pressure of the cooler, denser air outside.

... and so to bicycle pumps

You may have noticed that when you are pumping up a bicycle tyre, especially as it becomes more highly inflated, the pump gets hot. (You can quickly demonstrate this by putting your thumb over the outlet of the pump and compressing the pump as hard as you can. The sides of the pump rapidly heat up. Incidentally, do not try this with any form of mechanically driven pump as a very nasty injury could result.)

If you think about it, you will see that the reason for this is the converse of the hot air balloon. When you compress a gas, the molecules within it are forced more tightly together and collide more violently. Because the molecules suddenly do not have the 'elbow room' that their energy level requires the gas has to give up energy and will do so in the form of heat. (For what happens when you suddenly decompress a gas see *Why does ice form in a carburettor even when the air temperature is several degrees above freezing point?*)

If you look at the type of air compressor frequently used to drive pneumatic drills or 'road rippers', you will see that the cylinders in which the air is compressed are very heavily finned to assist cooling by presenting a greater surface area to the atmosphere. If the compressor were not cooled, it would rapidly become inefficient as a result of pumping a constant *volume* but decreasing *mass* of air to the drill. For the same reason, the part of a car's supercharger or turbocharger where the mixture is compressed is heavily cooled. (See *Why does a car seem to perform better on a cold, damp day?*)

However, the heating of air under compression can be put to good use. For example, in a Diesel engine, which is a form of compression-ignition engine named after Rudolf Diesel (although he was far from being the first to suggest the idea), the heat of compression within the engine cylinder(s) ignites the fuel and sparking plugs can be dispensed with.

An historical aside

It is interesting to note that while people in Europe were fiddling around with flint and tinder, their Oriental contemporaries had developed an effective form of lighter which used compression ignition. Typically this consisted of a short air-tight wooden cylinder with a closely-fitting piston moving within it. A piece of tinder was inserted in a small recess on the face of the piston. When the air in the cylinder was compressed with a sharp blow from the fist, the heat generated was sufficient to ignite the tinder. These 'fire pistons' were known in Burma, Indonesia, and Malaya where the name for the device is *gobek api*.

Although widespread in the East, there are only scant records of such lighters in Europe. There is some evidence of them in France and Switzerland while in England there is a record of a Mr James Burch Walker of Bromley in Kent who in the early nineteenth century had a walking cane which used the fire piston principle. He is reputed to have used it to light his church-warden pipe.

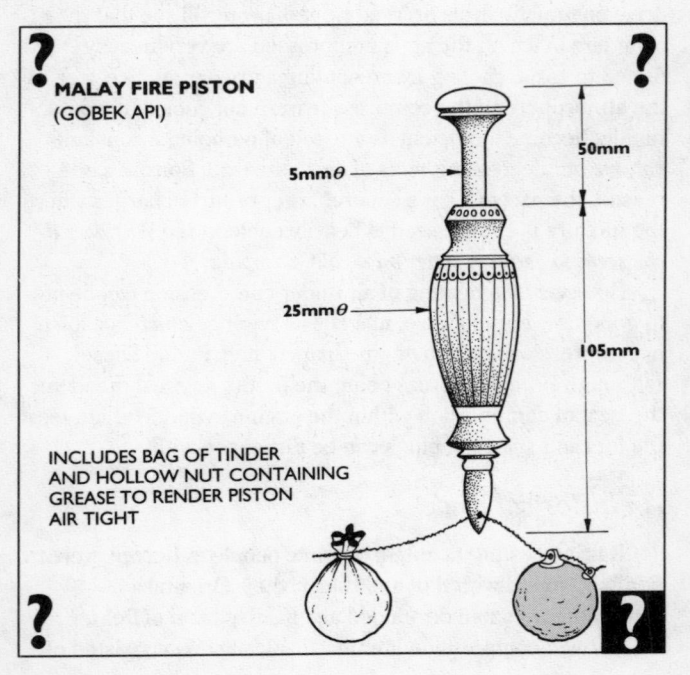

MALAY FIRE PISTON
(GOBEK API)

5mm θ

50mm

25mm θ

105mm

INCLUDES BAG OF TINDER
AND HOLLOW NUT CONTAINING
GREASE TO RENDER PISTON
AIR TIGHT

Why is a light material such as Kevlar well-nigh bullet-proof ❓

And when she could no longer hide him, she took for him an ark of bulrushes, and daubed it with slime and with pitch, and put the child therein; and she laid it in the flags by the river's brink.
Exodus, Chapter 2, Verse 3

We are not the first to cite the Book of Exodus in this context - there is a famous labour dispute in Chapter 5 concerning the manufacture of bricks without straw which is sometimes touched upon - but you can see from the above quotation that Moses had an even earlier exposure to what we would now refer to as 'composites'. (The translators of the King James version of the Bible used the homely term 'bulrush' for what was more likely to have been the papyrus reed, *Cyperus papyrus*.)

A composite material is generally one in which strands of fibre (bulrushes) are mixed with another material called the matrix (slime and pitch), in order to reinforce it. Examples include *papier mâché* (which was used by the ancient Egyptians in the manufacture of mummy cases and more recently in the former East Germany for the body panels of the Trabant motor car), Bakelite mouldings and also resins reinforced with glass or carbon fibres. In some senses reinforced concrete - in which iron rods are used to compensate for the weakness of concrete when it is under tension - is a composite, while wood, sometimes referred to by engineers as 'nature's composite', consists of a resinous substance called lignin reinforced with cellulose fibre.

'Kevlar' is the trade name for a synthetic fibre created by

Stephanie Kwolek, a Du Pont research scientist, in 1965. Basically it consists of man-made chain molecules known as polymers and depends on the very strong bonds between rings of carbon atoms for its incredible tensile strength (about five times that of steel, weight for weight) and minimal stretchability. At the moment it probably represents the state of the art in man-made 'wonder' materials.

The behaviour of composites is highly complex and is not to this day fully understood, and their use in engineering structures continues to be regarded as one of the 'black arts'. Also, depending on the material of the reinforcing fibre - whether it is chopped or long, random or woven, flexible or stiff, stretchable or not stretchable - there are probably as many different properties of composites as there are composite materials. Large books have been written on single tiny aspects of this subject, so what follows involves considerable over-simplification.

OK, so let's get down to it

OK. All surfaces have an inherent energy. The most readily-observable example of this is in the surface-tension of water. If you watch an insect walking on water you will see that its legs create dimples in the surface of the water. These, obviously, increase the area of the surface. The insect is supported because its potential energy (as a result of its weight) is exactly balanced by the increase in surface energy.

Liquids tend to reduce their surface energy as much as they can. This can be seen when a narrow stream of water from a tap breaks up into individual drips. The point at which this happens is when the total area of the surface of the drips is less than that of a thin cylindrical column containing the same volume of water.

In fact, solids behave in exactly the same way and the creation of a greater surface area in a solid will always require an input of energy.

The bullet

The speed of a shock wave in any substance is normally the speed of sound in that substance - which is going to be considerably faster than the speed of a bullet. Consequently, when a bullet hits a solid the shock waves will travel outwards from the point of impact at a far higher speed than that of the bullet itself. This means that a shield should have plenty of time in hand to do its job. In a fibre composite the shock will spread as a series of cracks in the matrix (the slime and pitch) which will tend to run along the surfaces of the fibres wrenching them away from the matrix material. In the case of Kevlar the fibres are so inelastic that very considerable forces are required to achieve this. If the fibres are randomly arranged in a criss-cross fashion, the cracks will branch and have to create even more

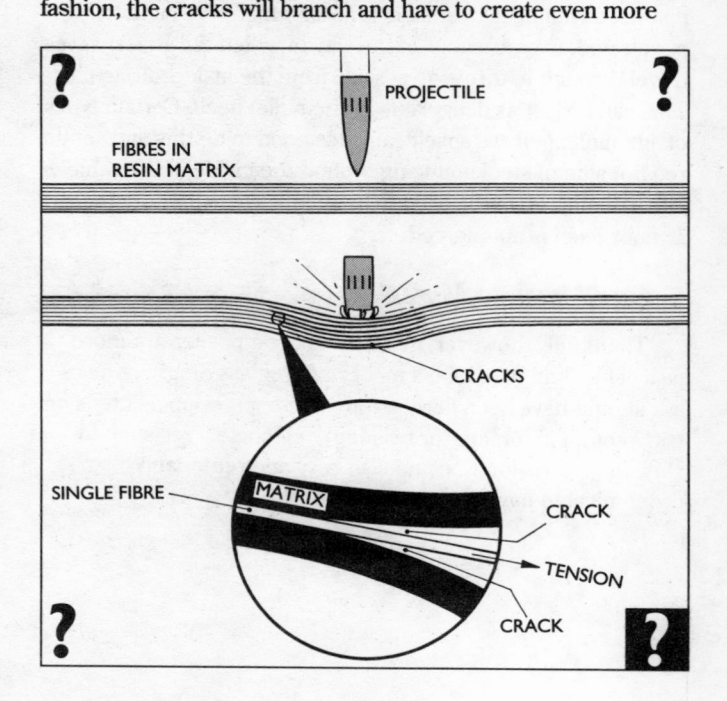

new surfaces every time they come to an intersection. A large crack does not have the chance to 'get going'. (This phenomenon is related to the reason it is so difficult to tear-up a telephone directory where the tear is 'blunted' and has to start again every time it reaches a new page. It also lends some credence to stories of lives being saved by a Bible in a battledress pocket. However, we digress . .) All of this happens in less than a hundredth of a second but the energy absorbed in the creation of all these new surfaces along the fibres can be quite enough to stop a bullet.

Why doesn't steel offer such good protection?

Apart from its light weight, this type of protection can have distinct advantages over steel armour which, even if not penetrated, may, because of the way in which the shock waves travel through it, throw off a 'scab' from the inside surface. This can be just as dangerous as the bullet itself. Certain types of anti-tank shell are specifically designed to do this and send a red-hot slug of steel bouncing around the inside of the vehicle like a demented high-speed billiard ball, even although the armour itself is not pierced.

. . . and back to Moses

Thankfully, however, most uses of composites are more peaceable than this. Like a modern fibreglass dinghy, Moses' ark should have been fairly strong in compression and tension, stiff enough in torsion (or twisting), adequately resistant to abrasion, corrosion-free and also easy and reasonably economical to manufacture as a one-off.

!

Why is a sixteen-valve engine regarded as a 'good thing' ?

Many European and Japanese car manufacturers make great marketing mileage out of the fact that their vehicles have a sixteen-valve engine. Or in the case of a six-cylinder car, a twenty-four valve and so on. In most cases the engines have four valves - two inlet and two exhaust - per cylinder. However, specialist engines with two inlet and one exhaust and even with three inlet and one exhaust are not unknown.

Why this seemingly needless complexity?

There is a general misconception that cars run on petrol. In fact they run on a mixture of atmospheric oxygen and petrol. The proportion of petrol in the mixture is seldom more than one fourteenth of the whole.

The greatest factor governing the output of an engine is the mass of fuel-air mixture that can be sucked into it, efficiently burnt and then blown out. Almost every aspect of 'tuning' an engine - from the turbocharger to polished inlet ports and extractor exhaust manifolds - seeks to further this aim. This factor is known as the volumetric efficiency of the engine - or its ability to 'breathe' well.

Try running a kilometre with your mouth closed

The functioning of a car's engine is, in many ways, like that of the human body. If you are exerting yourself - say running a kilometre - one of the greatest limitations on your performance is the rate at which oxygen can be drawn into the lungs and made available to oxygenate the blood supply. In an engine, the intake of 'breath' is an equally limiting factor.

If you look at the diagrams below, you will see that this is better achieved by a greater number of small valves than by a lesser number of big ones.

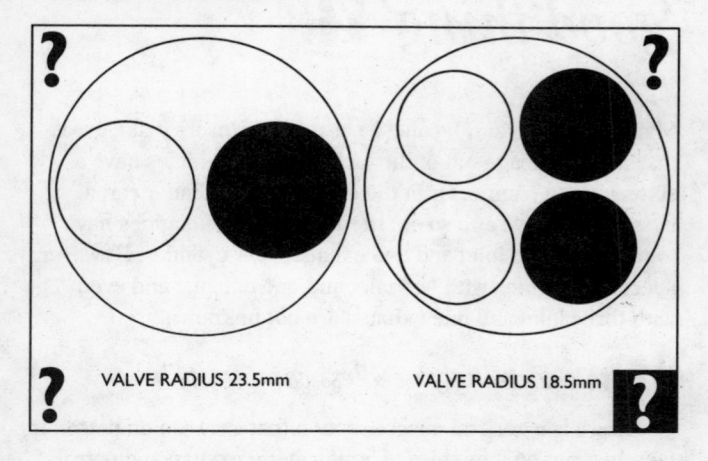

VALVE RADIUS 23.5mm   VALVE RADIUS 18.5mm

In the first diagram there is one inlet and one exhaust valve which are both as large as they possibly can be, within the confines of the combustion chamber.

In the second diagram there are four valves - two inlet and two exhaust - each as large as they can be, again, within the confines of the combustion chamber.

To go through the mathematics - the area contained by a circle is expressed by the formula π multiplied by the square of the radius of the circle r. In the first diagram the radius of each valve is 23.5mm. So -

$$\pi \times (23.5 \times 23.5) = 1735 \text{ square millimetres}$$

In the second diagram, each of the four valves has a radius of only 18.5mm. Thus each valve has an area of -

$$\pi \times (18.5 \times 18.5) = 1075 \text{ square millimetres}$$

Consequently, the two-valve head has a total 'breathing area' of 1735 square millimetres on the inlet side, while the four-valve head has an equivalent of 2 x 1075 or 2150 square millimetres.

In fact there are many different considerations in the design of a combustion chamber such as controlling the position of the combustion and generating turbulence in the fuel-air mixture, which, among other things, generally helps it to burn more evenly.There is also the practical consideration of complexity - of having to 'get it all in' with the other parts of the engine such as the camshafts - as well as the mechanical considerations of the friction and inertia of the different components. However, the general principle that it is easier to maximise breathing area by using many small valves rather than fewer big ones, holds good.

You might think 'Why stop at four valves?' and you would be absolutely correct. There are several racing engines with five or even six valves. At the moment the ultimate in this line of development - at least as far as production engines are concerned - is an engine developed by Honda for the NR750 motor bike. This has eight valves per cylinder and requires oval pistons and cylinders in order to fit them all in.

!

Why is an owl able to fly so silently ❓

We humans have often regarded the owl as a symbol of wisdom. Many bird-watchers might argue that, when compared with the low cunning of the crow, the owl is actually rather dim. However, the owl does have a unique advantage amongst birds in the design of its wings.

Most birds generate a rustling noise as they fly. This is caused by aerodynamic turbulence in the form of small eddies along the top of the wing and in frictional noise as the feathers slide over each other.

The primary flight feathers (those on the leading edge towards the outer portion of the wing) of all species of owl exhibit three unique features:

1. The upper surface of each feather is covered in a soft down. This damps out turbulence as air flows over the top of the wing. In addition, it reduces the frictional noise made by the feathers when they slide over each other as the owl adjusts the wing shape when manoeuvring.

2. The leading edge has a comb-like structure that reduces flow separation and consequently further reduces noise.

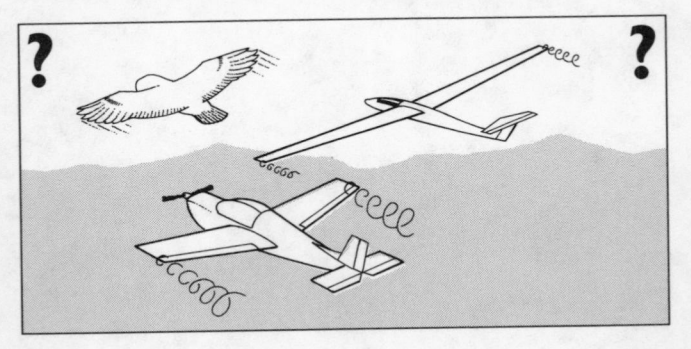

3. The trailing edge of each feather is slightly shaggy. This has the effect of further damping out airflow noise as the air leaves the wing, especially at the wing tip vortices. (Other examples of shaggy material being used to damp out turbulent energy, include the shaggy microphone guards - called 'Dougals' after the dog in 'The Magic Roundabout' - used by sound recordists on outside broadcast teams on a windy day and the brush-like material on the edge of the mudflaps of heavy lorries which go some way towards breaking up the force of road spray.)

Whenever air-flow noise is apparent it follows that turbulence is present. Turbulence absorbs energy and therefore silent flight is efficient flight.

However, in the case of the owl, one suspects that the main advantage, when coupled with sharp vision and acute hearing, is not in energy saving but in allowing it to swoop silently on unsuspecting prey.

Why are maps drawn with North at the top ❓

Nowadays, it is hard to visualise a map that does not feature north at the top, but this was not always so.

In a sense, the earliest maps were cave drawings depicting primitive man in pursuit of game, sometimes with representations of topographical and celestial features included. Their message, apart from any ritual significance they may have had, was simple: here, and at this time of the year, there is food to be found.

The oldest known map in the accepted sense of the word was drawn on a clay tablet about 3800 BC, and depicts the river Euphrates flowing through northern Mesopotamia, Iraq. This, and others that followed it, were little more than rough sketches of localised features; it was not until many centuries later that the astronomical and mathematical expertise of the ancient Greeks placed cartography - the science of map-making - on a sound footing.

At the forefront of the pioneers in the field was the Greek mathematician and philosopher Claudius Ptolemaeus (c. AD 90 – 168), more popularly known to history as Ptolemy. The last great scientist of the classical period, he was the first to draw a map of the civilised world that was based on all available knowledge, rather than conjecture. Earlier, the Babylonians had attempted to map the world, but they presented it in the form of a flattened disc rather than a sphere, which was the form adopted by Ptolemy.

Inevitably, given the state of knowledge of those times, he got things wrong; for example, he showed the British Isles with the territories of what are now England and Scotland joined back to front, and his estimate of China and the Atlantic Ocean were wildly inaccurate. Nevertheless, it was a laudable effort, and the map – together with an eight-volume Guide to Geography, in which he summarised the work of all earlier scientists and geographers – remained a work of reference for over a thousand years. In fact, Christopher Columbus used a version of it when he set sail in search of the New World - which caused him some navigational problems, since Ptolemy had miscalculated the size of the Atlantic and was unaware that the Pacific Ocean existed.

The really important thing about Ptolemy's map was that north was at the top. The reason for this was that he decided to orientate the map in the direction of the Pole Star – a perfectly sensible thing to do, since Polaris was the immovable guiding light in which the voyagers of that era placed their trust.

North at the top remained the accepted arrangement until the early Middle Ages, when religious dogma began to interfere seriously with the advance of science. In accordance with the dictates of the Church, maps were still produced in accordance with Ptolemy's principles – but now Jerusalem was the central feature, as it was held to be the centre of the Christian faith, and *east* was moved to the top.

These representations are often called 'T' Maps because they show only three continents - Europe, Asia and Africa - separated by the 'T' formed by the Mediterranean Sea and the River Nile. From a navigational point of view, they were virtually useless.

More accurate maps began to appear in the 14th century, with the spread of maritime trade and exploration and increasing reliance on the magnetic compass, a device first used in primitive form by the Vikings. Once again, north assumed its rightful place at the top of maritime charts, so that features could be readily aligned with true and magnetic north. Its position there was permanently fixed in 1569 when the foremost cartographer of the Age of Discovery, Gerardus Mercator of Flanders, developed a cylindrical projection, intersected by north-south lines, that represented the curved surface of the Earth on a flat map.

Why does a glider have such long, tapering wings when compared with a powered aircraft ?

In designing a glider one of the most important considerations is that it will fly as far as possible from a given height before it has to land. To achieve this it is essential to prevent energy from being wasted in the form of air resistance or drag.

Unfortunately one of the trade-offs for increased lift is increased drag. A powered aircraft simply punches its way through this problem with its engine, but what can a glider do?

Lift is created as the wings move through the air, and relies on the fact that air pressure drops as air speed increases. Since the upper surface of a wing has a greater curvature and surface area than the lower, the air which is forced to 'go the long way round' over the upper wing in the same period of time (ie faster) produces a drop in pressure on the upper surface of the wing – ie lift.

? **?**

LIFT

LOW PRESSURE

HIGH PRESSURE

At the wing tips there is a tendency for the positive-pressure air under the wings to seek out the negative-pressure area on the wing top. This causes what is known as a wing-tip vortex. The bigger these turbulent areas the greater the energy lost in drag.

If the wings are short and stubby, the energy lost through these vortices would be considerable. If, however, the wings are long and taper to a small point, the vortices can be kept to a minimum and hence maximise glide efficiency.

A recent development in the pursuit of fuel efficiency has been to fit vertical 'winglets' to the end of aircraft wings. These help prevent the migration of air from the lower to the upper wing and 'unwind' the vortex turning it into useful lift.

The simplest explanation can be found in the diagram below:

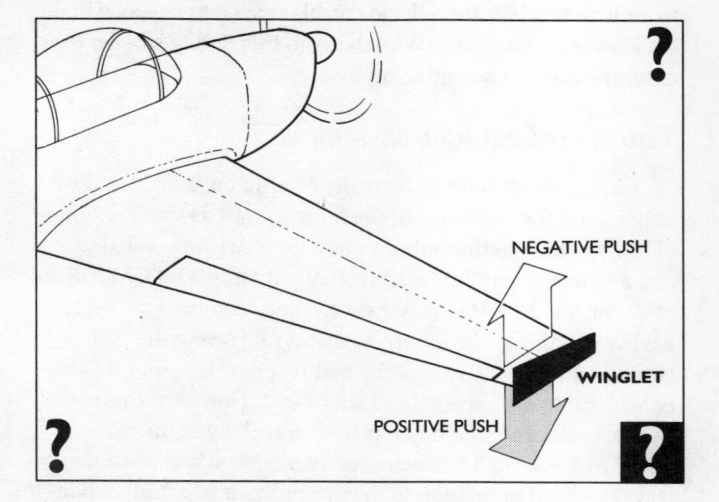

(See also *Why is an owl able to fly so silently?*)

Why does bathwater go down the plughole one way in the Northern Hemisphere and the other way in the Southern ?

Despite various stunts at Equatorial crossing points on the roads of Africa, which claim to 'demonstrate' this phenomenon to gullible tourists, there is no earthly reason to suppose that it could ever be observed. Why then did this dubious belief ever enter the popular scientific mythology?

Trains and planes and... guns

First of all we have to be really clear in our minds that we understand the workings of the force known as momentum. (This is not altogether intuitive and even Aristotle failed to come to terms with it.) Imagine that you drop a billiard ball out of a train window. It is very easy to allow 'common sense' to lead you to think that the train rushes on leaving the ball behind while it falls so that the ball reaches the ground some considerable distance behind your hand. However, this is not the case. Because the ball has been travelling at the same speed as the train it has acquired momentum and continues to travel forward in relation to the ground as it falls and, although it will lose a tiny amount of speed because of air resistance, will hit the ground almost perpendicularly underneath your hand – just as it would have if you had dropped it inside the carriage.

Similarly, if you have ever lain on a beach and idly

wondered what would happen if the undercarriage were to fall off an overhead airliner, there is no use consoling yourself in the belief that by the time the plane is overhead the undercarriage must have safely hit the ground several kilometres back. Provided the plane is travelling at a constant speed, although the undercarriage actually parted company with the aircraft several kilometres back, it will hit you at precisely the moment the plane appears overhead. This will happen whatever the altitude and speed of the aircraft, provided they are both constant – although, again, ignoring the effects of air resistance.

... *and guns*

Now imagine a hypothetical (and somewhat uninformed) gunner sitting in London who is ordered to 'take out' Edinburgh Castle, some 600 kilometres to the North. He looks at his map, aligns his gun on the seemingly correct bearing and sets up an appropriate trajectory so that the shells should fall exactly on his target. Imagine his surprise when he hears reports that the shells are falling harmlessly in the Firth of Forth, several kilometres to the east.

The reason for his puzzlement is that he has neglected to take into account the fact that London, because it sits on a latitude with a greater radius in relation to the Earth's axis of spin than Edinburgh's, is spinning (eastwards) faster than the target. When the shell is fired it has a momentum caused by this speed and just like the billiard ball dropped from the train continues to travel in an easterly direction with this momentum. An equally uninformed gunner retaliating from Edinburgh will be equally surprised when his shells mysteriously fall somewhere west of his target. (In terms of billiard balls this is equivalent of dropping one from a bridge into a railway truck. The ball has no lateral momentum and consequently rolls to the back of the truck.) There is a general rule, therefore, that objects moving in any direction in the Northern Hemisphere will drift towards the right and in the mirror image of the Southern Hemisphere, towards the left. This is known as the Coriolis effect after Gaspard de Coriolis (1792 – 1843), the mathematician who first described it.

What's this got to do with bathplugs?

Bear with us a moment. It may be difficult to think of moving air as 'an object' but in fact it too is subject to the Coriolis effect. As the old mnemonic goes, 'Pressure will always flow from high pressure to low', as can all too easily be seen when a car tyre is punctured. It is sometimes convenient to think of this as a kind of 'pressure gradient' like say, a ball rolling down a slope.

The isobars depicted on a weather map are simply lines linking places which have the same pressure, exactly like the contour lines on an ordinary map which link places of the same height. From this you might assume that air would move down the pressure at right angles across the isobars in much the same way as things roll downwards. Not so. In fact the air moves almost in parallel with the isobars. To understand why, consider the diagram. Imagine a pocket of air between the two isobars shown. On the one hand there is a force pulling it down the pressure gradient across the isobars. However, the

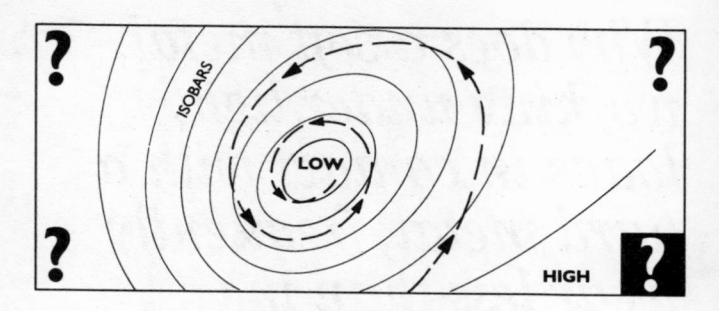

Coriolis effect is also acting on it, pulling it at right angles towards the right in the Northern Hemisphere. The two forces fall into an exact balance and the air forms an anti-clockwise spiral, known as a cyclone, around the area of low pressure. This will always happen in the Northern Hemisphere. In the Southern Hemisphere air moves around an area of low pressure in a clockwise direction but it is still known as a cyclone. If you imagine the reverse situation, with air moving away from an area of high pressure, you will see why the air moves clockwise in the Northern Hemisphere and anti-clockwise in the Southern. This is called an anticyclone in both cases.

And so, finally, to bathplugs

Because of this well-observed phenomenon in the weather, people began to look for it on a smaller scale and because water goes down the plug in a spiral this was one of the places they started looking. It is probably a fruitless search. The Coriolis effect only makes a significant difference on the large distances involved on the weather map and the way your bathwater runs away is far more likely to be influenced by local factors such as where the taps are, which side of the bath you got out of, dirt around the plughole or how you pulled the plug out. To date, the most sensitive experiments have failed to measure any evidence of the Coriolis effect on this scale. The myth, however, persists.

Why does a soft metal, working under light loads in contact with a hard metal, frequently wear less than the hard metal ❓

In the less than perfect conditions of the real world grit has a tendency to make its presence felt. Mineral grit, essentially tiny particles of silica, is not only one of the hardest materials but is, perversely enough, the combination of the two most abundant and widely-distributed elements on earth – silicon and oxygen. (When silicon combines with the third most abundant element, carbon, you really have trouble in the form of carborundum!) The grit will usually be harder than either metal. However, it will become firmly embedded in the softer metal in preference to the harder, forming a stable and highly abrasive surface.

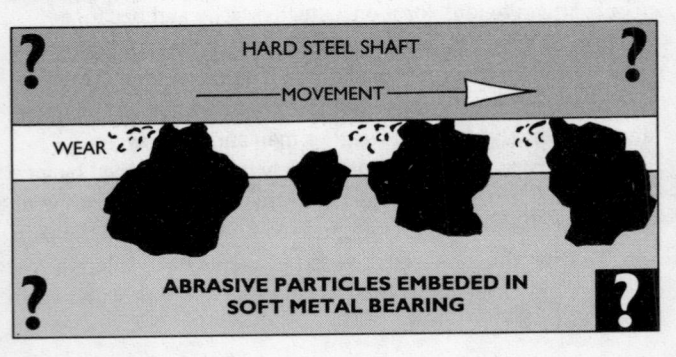

HARD STEEL SHAFT

MOVEMENT

WEAR

ABRASIVE PARTICLES EMBEDED IN
SOFT METAL BEARING

This phenomenon can often be observed in the float chamber of an ageing carburettor. A hard carbon steel needle slides in a brass female orifice to form a valve which regulates the supply of fuel. Paradoxically, it is always the needle which has to be replaced or re-machined first.

Similarly, in a washing machine driven by a rubber belt, the spindle will develop deep grooves because of the grit embedded in the surface of the belt, while the belt itself will remain unworn.

!

Why is all flesh grass ?

All flesh is grass and all the the goodliness thereof is as the flower of the field.
Isaiah, Chapter 40, Verse 6

While the writer of the Book of Isaiah was referring to the transient nature of life, there is another fundamental reason for thinking of flesh as grass. While carbohydrates and nitrogen – an essential constituent of proteins – are vital to all animals, despite the abundance of carbon, hydrogen, oxygen and nitrogen all around us, no animals are able to use these elements directly. Only plants can be considered as primary producers of food. Even if you are an enthusiastic carnivore, it is ultimately the grass (or barley) upon which the bullock fed which created the essential nutrients in your steak – not the bullock itself.

!

Why is electricity transmitted across the country at such a high voltage ❓

In a modern power station electricity is generating at around 25 000 volts of alternating current. The domestic user draws 240 volts from the mains. However, electricity is transmitted across country at voltages of around 400 000 volts or more. Why?

Since the electrical energy transmitted in a given time is the product of the current and the voltage, you have the choice of either transmitting the same amount of energy at a high current and low voltage or at a low current and high voltage.

So, which is better?

There are several advantages in going for the low current - high voltage option. A low current has a much smaller heating effect in the transmission cables so less energy is lost to the atmosphere in the form of heat. Also, thinner wires can be used which means that the supporting pylons can be spaced out over greater distances so that fewer are needed. (Electrical transmission lines are generally made of aluminium with a galvanised steel core. This is because aluminium is very light for a metal and is, weight for weight, an excellent conductor of electricity - most of the current in a wire is conducted along the surface. However, it is not strong enough to support itself over the distances required and so a steel core is introduced to give additional strength.)

The disadvantage of using a high voltage is the danger of current arcing to other transmission lines or the pylons, so the transmission lines have to be kept well away from each other and greater insulation has to be used at the pylons. It is also the reason that pylons have to be so high in order to keep the lines well away from the ground.

!

Why do airlines prohibit you from carrying thermometers ❓

If you read the list of prohibited items printed on your airline ticket you may be surprised to see such an innocuous-seeming item as an ordinary mercury thermometer included along with the apparently more obvious items such as firearms.

Mercury and some metals such as the aluminium alloys which are used in aircraft construction react powerfully together to form a special type of alloy known as an amalgam. One can sympathise with the aircraft operator wanting to avoid the risk of corrosion - many more hours of work are spent on corrosion rectification and avoidance on an aeroplane than on purely mechanical problems – but isn't the warning rather extreme?

Anyone who has ever seen mercury reacting with aluminium will think not. It is a violent and rapid reaction which can destroy a piece of aluminium alloy sheet in minutes producing (rather beautiful, if one's life is not at risk) feathery fronds of aluminium amalgam. The mere suspicion that there was a loose thermometer on board would be grounds for a pilot to refuse to take off. If you do have to carry items containing mercury on an air journey you should warn the airline well in advance and they will make provision for it as a dangerous cargo.

Why are eggs egg-shaped ❓

We normally think of bird eggs as rather brittle objects. In fact, weight-for-weight, eggshells are incredibly strong. There is a report in *The Guinness Book of Records* of an ordinary domestic hen's egg surviving a fall of 198 metres from a helicopter over a Japanese golf course in 1979. This is not as unlikely as it might seem and if you throw an egg as high in the air as you can over a grass surface it will almost invariably survive the landing intact.

Eggshells are made of calcite, a form of calcium carbonate, with the individual crystals arranged in an interlocking pattern in such a way that, rather like the stones in the arch of a bridge, external pressure forces the crystals together to strengthen the structure. This arrangement is weak, however, when confronted with forces from inside – which is just as well for the chick when the time comes to hatch out. Part of the calcium required for making the shell comes directly from the mother's diet, which is why it is usual to feed ground-up eggshells to free-range domestic hens. Other species adopt different strategies to supplement their calcium intake during

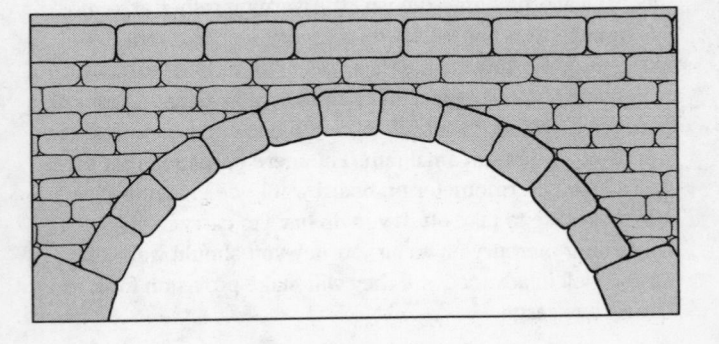

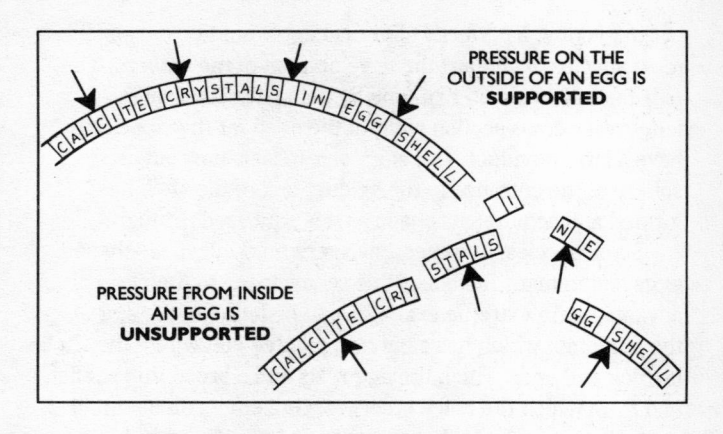

PRESSURE ON THE OUTSIDE OF AN EGG IS **SUPPORTED**

CALCITE CRYSTALS IN EGG SHELL

PRESSURE FROM INSIDE AN EGG IS **UNSUPPORTED**

CALCITE CRYSTALS IN EGG SHELL

the breeding season. For example, the females of big, carrion-eating species such as vultures, which generally stick to a diet of pure meat torn from large carcasses, will switch to killing small animals and eating them whole in order to derive the benefit from the calcium in the bones. Sandpipers in the Arctic, normally insect eaters, have been known to pick at the skeletons of long-dead lemmings for the same reason.

However, the mother bird is never able to absorb enough calcium from her diet during the formation of the egg(s) and consequently has to sacrifice calcium from her own bones. Birds have developed special medullary bones in which the female lays down extra reserves of calcium in the marrow during the weeks leading up to laying. After laying, there can often be more calcium in a clutch of eggs than there is in the entire skeleton of the mother. You can see that the business of making eggshells is an expensive one in terms of the hen's feeding and metabolism.

In a structural sense (and in the sense of economic use of materials) the best shape for an egg would be a perfect sphere with the stresses evenly distributed throughout a uniformly thick shell. Indeed some eggs, such as those of the tawny owl, very nearly approximate to this ideal, but they are the exception rather than the rule. Why?

In general, a perfectly spherical egg would be too small in most species to support the development of the embryo. The egg can have a greater volume by being slightly oval in lengthwise cross-section without the need for that species to have a larger oviduct. The egg does, in fact, start out as a sphere at the entrance to the oviduct before the shell has formed and acquires its shape as it is squeezed through.

Some species, however, have very markedly pear-shaped eggs and in many instances this seems to act to their advantage. An extreme example is to be found in the egg of the guillemot which has a narrow, pointed end where the shell is thick and upon which the egg rests and a broad thin-shelled end from which the chick emerges. Guillemots do not build nests but lay their single egg on the ledges of exposed and often storm-wracked cliffs. Because of the great difference in radius at the two ends, a guillemot's egg will not roll in a straight line but instead describes a fairly tight circle of about 7 inches (18 centimetres) radius when it is newly laid and, because of a slight change in the egg's centre of gravity as the embryo develops, about 4 inches (10 centimetres) radius when it is fully incubated. Although many guillemot's eggs do roll off cliffs in gusts of wind it is reasonable to assume that many more would be lost in this way if it were not for their distinctive shape. A domestic chicken's egg in contrast will roll for some distance in an almost straight line.

Another species to have a markedly pear-shaped, indeed almost conical, egg is the lapwing, but here the advantage is of a totally different sort. In order for the embryo to develop, all

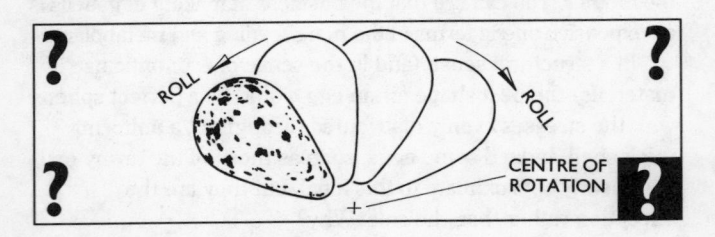

ROLL

ROLL

CENTRE OF ROTATION

birds' eggs have to be maintained at the blood temperature of the adult. This is achieved by one of the parents sitting on the egg or clutch of eggs. As with most small birds, the egg of the lapwing is fairly large in relation to the size of the adult. Lapwings almost always lay clutches of four eggs and the shape of the eggs allow them to 'pack' very efficiently in the nest with the narrow ends pointed inwards under the incubating parent who is thus able to more effectively cover the clutch and impart a greater amount of heat than would be possible if the eggs were more spherical.

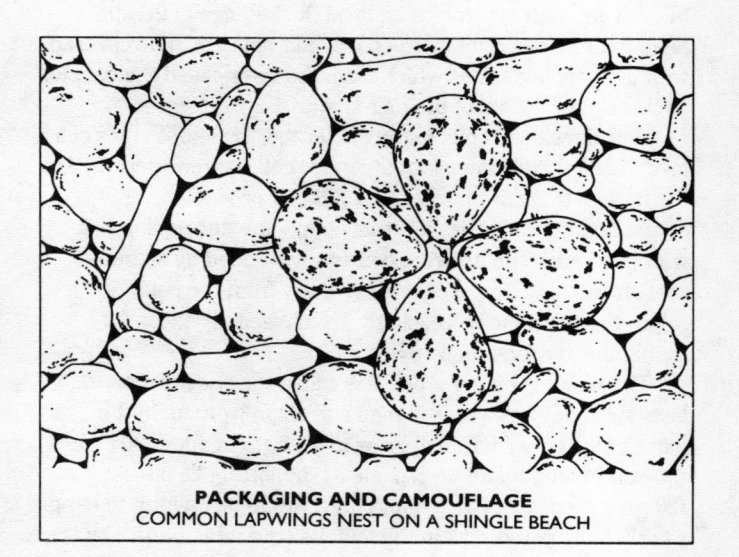

PACKAGING AND CAMOUFLAGE
COMMON LAPWINGS NEST ON A SHINGLE BEACH

Why does hair turn grey ❓

If you could come up with a complete answer to this question, you could end up very wealthy. Scientists known why hair turns grey, but they do not yet know what triggers off the action.

Hair is very complex, and it is only comparatively recently that careful research has disentangled its structure. In essence, a hair is a thin fibre composed mainly of proteins called keratins. All skin is keratinized, and hair, nails, horn and hooves are nothing more than local thickenings of keratin. Seen under a powerful microscope, hair appears to be covered with overlapping scales which form the layer called the cuticle, enclosing the cortex of the hair shaft.

Hair develops in pits in the epidermis, the skin's outer cell layer, and extends deep into the inner cell layer, or dermis. At the base of the hair follicle is a papilla, or projection, containing the blood vessels that supply the developing hair with nutrients. The growing hair is surrounded by an inner root sheath, and as the cells move away from the point of origin they become increasingly keratinized, drying, hardening and eventually dying.

The hair – growing process is called the anagenic phase; it lasts for about four years in men and six in women. In the latter case, a head hair will grow to a length of about 30 inches (80 cm) in length and be capable of supporting 28 ounces (80 g); a thousand, twisted together, would be enough to hang a fairly hefty person(as in the famous story of Rapunzel by the Brothers Grimm). The growing process is not continuous; after its initial activity the follicle enters a resting phase lasting from three to six months before it starts work again, forming a new hair to push the old one out. About 100 hairs are shed each day as normal turnover, and at any one time about 90% of the hairs are actively growing.

Hair colour is due to a pigment called melanin, (see *Why does skin pigmentation vary?)* which is produced by cells called melanocytes. Melanin itself is brown; whether a person's hair is blond or very dark depends on the amount of melanin produced, and how it is distributed. Red hair contains an extra pigment, rich in iron.

It is when the melanocytes cease to function that hair loses its colour. In fact, there is no such thing as grey hair; it turns white. The grey appearance is caused by the intermingling of white hairs with those still retaining their colour.

Exactly why melanocytes give up the ghost is still something of a mystery. In the case of old age, it is logical to assume that they slow down in keeping with the rest of the body's metabolism. But white hair can be hereditary, and melanocytes can also lose their natural function as the result of shock, stress or anxiety. Cases where a person literally 'turned grey overnight' are well documented.

!

Why is it dangerous to use a charcoal-fired barbecue in a confined space **?**

One of the products of the incomplete combustion of charcoal in air is carbon monoxide. This is an extremely poisonous gas and air containing as little as 0.1% (by volume) could easily be fatal if continuously breathed for more than a few minutes. It is lighter than air and will form lethal pockets if trapped beneath the low roof of a garden shed or car port. (This is one of the reasons why, if you are ever unfortunate enough to have to escape from a smoke filled building, you should crawl as quickly as possible on your hands and knees.) It is particularly dangerous in that it has no smell.

The reason for its very considerable toxicity is that it has a great affinity for, and forms a very stable bond with, the blood's haemoglobin. This is the substance which carries oxygen around the body from the lungs so that it can oxidise food substances in order to release energy. The bond between oxygen and haemoglobin is a relatively weak one so that the blood gives up its oxygen fairly easily. The bond with carbon monoxide, however, is a much stronger one and once it is formed, the carbon monoxide cannot be replaced by atmospheric oxygen. Therefore, once carbon monoxide poisoning has set in, it rapidly becomes an irreversible process.

Normally, one would expect someone suffering from any form of asphyxia to have a bluish colour showing through the skin as this is the colour of de-oxygenated blood. Someone suffering from carbon monoxide poisoning, however, goes a vivid pink as the compound of carbon monoxide and haemoglobin is a bright scarlet. Even after death the haemoglobin retains the bright red colour of arterial blood.

Carbon monoxide is present in the exhaust gases of petrol engines and most people are aware of the dangers of breathing these fumes. However, unlike the canny charcoal burners of old, people are not now so wary of the temptation to move a charcoal fire under shelter.

Why is it advisable for electrical transmission lines to cross steel fences at right angles ❓

It is generally more efficient to transmit electrical energy at high voltage and low current (See *Why is electricity transmitted across country at such a high voltage?*). Any change in current will induce an opposite (but not equal) current in a parallel conductor. Since the current in the transmission line is an alternating current, typically of 400 000 volts, running at 50 Hertz – ie changing direction 50 times a second – it will induce a significant current in any convenient parallel conductor, such as a steel fence, even though they are not in any way directly connected. The nearer the fence is to the transmission line and the longer it is, the greater the effect and the danger.

Should a farmer be unwise enough to put up a long wire fence that runs parallel to a power line then the result could be some nasty shocks. However, if the fence crosses the path of the transmission line as near to a right angle as possible, the current induced in the fence will be negligible.

Why is antimony always a constituent part of the alloy used in casting type ❓

Type metal is normally an alloy of 60% lead, 30% antimony - a brittle metallic element - and 10% tin. While antimony adds hardness to the alloy, the need for this was surpassed when pages of type set on Monotype or Linotype machines were merely used as patrices to make moulds from which the actual printing plates were formed. The metal plates were immediately melted down and their durability was not an issue. However, antimony continued to be used in type metal. Why?

Like water (see *Why does ice form on the top of ponds?*), antimony has the most unusual property of expanding when it cools and solidifies. Bismuth is the only other element to behave in this way. This means that when it solidifies in a casting matrix the expansion of the metal forces it to fill every nook and cranny of the mould in a way that would be impossible even with the most extreme external pressure. Since a serif - the flat pointed line at the at the end of the strokes in a letter - is regarded as an important ingredient in the legibility and elegance of many type faces, the presence of antimony is essential to ensure that these are sharply defined.

X

Bismuth could, in theory, also be used and indeed has been in the past - but since it imparts very low melting points to its alloys this would slow the casting process down while the type caster waited for the type to solidify. Also, iron or steel with a high graphite content displays a similar expansion when it solidifies and is sometimes used in very intricate ornamental castings. However, its relatively high melting point - certainly over 1000 degrees Celsius — makes it inconvenient (and out of technological reach, in terms of heat supply, until the late nineteenth century) in a type foundry, especially when the strength of steel is not required.

Pure antimony has no (as yet discovered) practical uses, but because of the above property it is very useful in alloys used in fine castings. It is also added in small quantities to lead when extra strength or hardness is required, as in car batteries and bullets.

!

Why does a car seem to perform better on a cold or damp day ❓

A car is travelling along a level motorway on a warm day. Suddenly it moves into an area of cold, damp air. The throttle opening remains the same yet the car goes faster. Why?

The fact is that it genuinely does perform better. There is a general misconception that cars run on petrol. In fact they run on a mixture of atmospheric oxygen and petrol where the proportion of petrol in the mixture is seldom more than one fourteenth of the whole.

The most important thing governing the output of an engine is the efficiency of its 'breathing'. (See *Why is a sixteen-valve engine regarded as a 'good thing'?*) Almost every aspect of 'tuning' an engine – from the turbocharger to the extractor exhaust manifold and polished inlet ports – seeks to further this aim. This factor is known as the volumetric efficiency of the engine.

Cool air is denser than warm air (See *Why does a hot air balloon rise?*) When the car enters the patch of cool air, the number of oxygen molecules entering the combustion chamber is significantly higher. In addition to this the water vapour in the damp air helps keep the mixture cool and provides a dense mass of gasses to be burnt. Consequently the force (combustive mass) of the explosion on the power stroke is greater, leading to an increased effective pressure on the piston crown and an obvious increase in power.

This is why a properly conducted test report of a car will always publish details of temperature and humidity when performance figures are being recorded. It is also one of the reasons that performance figures for the same car can vary so widely. The manufacturers when recording their own published figures doubtless choose a chilly damp day for the test runs.

!

Why does skin pigmentation vary ❓

No single issue in the history of mankind – except, perhaps, the obvious one of religion – has caused more controversy and incited more passions than the colour of a person's skin. And it all arises out of substances present in our bodies over which we have no control.

Skin colour, or pigmentation, ranges from the lightest Caucasian complexions of northern Europe to the darkest Negroid of central Africa. These are the two extremes, but in between lie the special shades of yellow-brown peoples such as the Chinese, Inuit and North American Indian.

Skin colour is the product of three pigment factors, all of which are present in every race. The first is melanin, the dark brown pigment granules in the cells of the epidermis, or outer layer of the skin; the second is melanoid, an allied dissolved substance also present in the epidermis; and the third is carotene, an orange pigment found in the outer layers of epidermal dead cells and the deeper fatty layer underneath the skin. Carotene, incidentally, is also found in carrots, tomatoes and various other coloured plants.

Just as is the case with hair colour, (see *Why does hair turn grey?*), variations in skin colour are caused by differences in the concentration of pigments present, especially melanin. Pigment in the outer epidermal layers also tends to block the reflection of blood colour through the skin, and variations in the colour of blood itself may have an effect too on the overall coloration of a person.

The wide variations in the skin colours of the human race are mainly the result of the evolutionary process, but hereditary factors also play a widespread role. In this respect the obvious factor is racial interbreeding between white and

black in which offspring show a blended coloration or a
resemblance to either parent, although albinism and
generalised skin colouring such as the presence of freckles are
also hereditary. Albinism is caused by a lack of melanin and,
although rare, is present in all races; affected individuals have a
tendency to suffer from skin cancers, because melanin acts as a
shield to absorb damaging ultra-violet radiation. Most fair-
skinned people are unable to produce enough melanin to
protect them from the effects of excessive exposure to sunlight,
which is why a `healthy tan' is really nothing of the sort.

In some races, the evolutionary process has led to the
production of just the right amount of melanin to match the
intensity of the sunlight that people are subjected to in their
everyday lives. In this context, the dark-skinned Inuit may
experience light that is just as intense as that produced by the
sun of Africa, even though the heat factor may be absent, for in
the polar regions light reflected from snowfields and ice floes
adds to the overall intensity.

Over the generations, the human race has developed an
exquisitely tuned response to the levels of ultra-violet radiation
that affects areas of the Earth in varying degrees. It is a
response that extends well beyond the production of extra
melanin to shield the outer layers of our skin, right through to
other protective measures that trigger our immune responses to
factors such as sudden climatic change. The intriguing question
is this: what would happen to our colour and appearance,
perhaps in just a few generations, if some climatic upheaval
reversed conditions in, say, Africa and northern Europe?

!

Why do the wagon wheels in movies sometimes seem to go backwards ❓

You may have noticed when watching a 'Western' that as a wagon or stage coach draws away from rest the wheels often seem to turn forwards, hesitate or 'wobble' in their apparent direction of motion and then turn backwards before the spokes settle down to a continuous blur of movement. A similar phenomenon can be seen in the televised view from the on-board camera in an open-wheeled racing car. Under braking, the lettering on the inside walls of the front tyres often seems to reverse its direction and move backwards. What causes the appearance of this impossible behaviour?

The moving picture of a film is, in fact, made up of a sequence of individual still pictures or 'frames' (usually 24 per second). In most circumstances our brain perceives – if you like 'kids itself' – this as continuous motion because it ignores the small gaps and discontinuities between the images and 'makes sense' of the images by running them together.

In certain circumstances, however, this effect breaks down – usually with regularly shaped objects in regular motion. Imagine that you are filming a perfectly regular spoked wheel which is turning at exactly 24 revolutions per second – this would be a very fast stage coach indeed, but please bear with us for the sake of argument. Each of the 24 still pictures or frames taken in that second will show the wheel in precisely the same position. Even though the stage coach will appear in the movie to be moving forward against the background, the wheels will appear to be absolutely stationary. If, however, the wheel turns at (say) 23 revolutions per second, it will just fail to make a complete revolution in each frame and, when the frames are strung together, will appear to be turning backwards.

What are harmonics?

The same thing can also happen at slower speeds if the wheels are turning at the number of revolutions per second of one of the factors of 24 – 12, 6, 4, 3, 1 – depending on the configuration of the spokes on the wheel. For example, in the case of a wheel turning at 12 revolutions per second each image shows a half revolution, but if the wheel has an even number of spokes and is perfectly symmetrical it will still appear to be stationary. If the wheel is turning at 3 revolutions per second, each frame will show one eighth of a revolution. As long as the configuration of the spokes is such that this gives an identical image (for example 8 or 16 symmetrical spokes) the wheel will again appear to be stationary. (We point this out to overcome any earlier objections you probably had about impossibly quick stage coaches. Equally, the same thing will happen at rates of revolution which are multiples, or harmonics, of 24 – 48, 96, 144, 192 and so on - or indeed at any rate of revolution which presents an identical image of the wheel in each frame.)

However, such regular and stable states of affairs are unlikely to last for long in the real world. Nevertheless, in moving off from rest the wheel has to pass through several points (numbers of revolutions per second) when this phenomenon will occur. If it is turning slightly slower than the rate of one of the factors of 24 it will just fail to complete an exact fraction of a revolution in each frame and again it will appear to be turning backwards. If the wheel is moving at a rate somewhere between two of these factors it will leave behind one state where it appears to be moving forwards and then, as it accelerates, 'pick-up' the next factor.But for a moment it will be moving too slowly to be exactly locked on to it and will appear to be moving backwards – hence the apparent 'wobble' as the wheel gathers speed.

Similarly, a television picture is 'refreshed' by the scanning beam 25 times a second. This is why it is difficult to take a photograph of a television picture. If the shutter speed of your camera is faster than one twenty-fifth of a second, the camera (which is not fooled in the same way as the eye) will 'freeze' a partially completed frame. Likewise, television pictures of other television screens or computer monitors always flicker because the scanning of the two images is slightly out of synch. If you have ever wondered when watching television features on the stock market why market makers in dealing rooms seem to put up with such dreadful equipment, the explanation is that they don't.

The above interesting but fairly trivial examples of this phenomenon do, however, point to some dangers (and practical uses) in real life.

Dangers ...

The light from a neon tube running on alternating current from mains electricity consists of a sequence of separate impulses of light. If the mains is running at 50 Hertz or cycles per second (as in the UK – similar values are common in most countries world-wide) there will be 100 such impulses each second since there is an impulse at each extreme of the cycle. In most circumstances, just as in watching a movie, the eye (or rather the brain) is unaware of this. However, in workshops and factories which are using rotating machinery, such as lathes, under this kind of light, the rotating parts can appear stationary when in fact they are rotating at 100 revolutions per second. It goes without saying that nasty accidents can result. (This problem is greatly diminished under ordinary light bulbs with a tungsten filament. As the filament continues to glow throughout the mains cycle, the effect of the impulses of light is smoothed out.)

... and practical uses - or turning the tables

This phenomenon can be put to good use, however. If, for example, you want to check or adjust the speed of a rotating object such as a record-player's turn-table you can do this with an instrument known as a stroboscope. This is simply a light which can be adjusted so as to emit pulses of light at a given frequency. A stroboscope flashing at a rate of 45 times a minute would make a turn-table rotating at 45 rpm appear stationary and therefore confirm that it was operating correctly.

!

Why is water wet ?

This may seem like a very odd question. We normally consider 'wetness' to be part-and-parcel of the properties of something being a liquid.

However, consider that you spilled water, aftershave, engine oil, polyeurethene glue or mercury over your hands, would you consider them all to be 'wet' in the same kind of way or to have the same degree of 'wetness'? (Incidentally we do not recommend that anyone actually tries this as mercury gives off an extremely toxic vapour and skin-contact with polyeurethene glue or engine oil is also best avoided.)

One of the things we are observing is the balance between the different properties of adhesion and cohesion. Adhesion is the attractive force between the molecules of the liquid and another substance, cohesion is a similar attractive force between molecules within the liquid.

If you push a glass tube into a beaker full of water, the water will rise up the tube - the narrower the tube the higher the water will rise. This is because the adhesion between the water molecules and the glass tube is greater than the cohesion within the water itself.

If the same experiment were to be carried out with mercury, the tube would cause a depression to form in the liquid. This is because the balance between adhesion and cohesion is the other way around. Also, for the same reasons, you will notice that in the beaker of water the edge of the liquid surface curves upwards along the sides, whereas with the mercury it curves downwards.

It is the adhesive properties of water which cause it to be soaked up by sponges, blotting paper and indeed your shirt in a rain storm.

Water is fairly adhesive but some other liquids show more extreme behaviour. The cooling system of a car can function perfectly satisfactorily until the addition of an ethylene glycol antifreeze when it mysteriously starts to spring leaks. This is because the new solution has a greater adhesion and 'seeks out' tiny gaps in joints through which it can creep. Other examples of liquids that display great adhesion are penetrating oil and brake fluid, which is why the joints in the hydraulic plumbing of a car have to be of a very high standard indeed.

Cohesion in water

One interesting example of *cohesion* in water is seen in the case of a traditional siphon of the sort used to empty a domestic fish tank.

At one time it was assumed that these worked entirely because of atmospheric pressure. Then it was discovered that – if all bubbles of air were scrupulously removed from the water – such siphons would still work in a vacuum! It had to be concluded that the cohesion in the water allows the longer column to pull through the shorter column because of the longer column's greater weight.

Why do birds sing the dawn chorus ?

Long ago it was widely believed that birds sang the dawn chorus as their expression of praise for the Creator, and for the enjoyment of mankind. Science, however, has destroyed that myth. We now know that the dawn chorus is an audible manifestation of aggression, sexuality and land-grabbing.

The mixture of sounds that make up the dawn chorus occurs in the breeding season and is produced largely by the males of the species. The range varies from the utterance of a single syllable to a complex and prolonged series of sounds, sometimes with a strong musical quality that makes birdsong so attractive to the human ear.

The onset of the dawn chorus in springtime is linked to the initial establishment of nesting territory, the male birds usually making themselves conspicuous by finding exposed perches in order to advertise their presence. Their songs reach a climax immediately before and during mating, then become less frequent when the young hatch and usually cease altogether when the chicks are fully fledged and the territory is abandoned. During the peak period, a male bird puts an enormous amount of effort into producing his song; one group of ornithologists, investigating the habits of a chaffinch, discovered that a single bird was capable of giving more than 2300 songs in one day.

Once the nesting territory has been established and an interested female arrives, the male bird intensifies his vocal and visual displays while he pursues her. This phase may last several days before the female becomes receptive and mating takes place; during this time the male may have to resort to physical violence to drive off rivals. Eventually, a 'pair bond' is created between male and female; in some cases it lasts only for one season, although many larger birds – swans, for example – remain paired for life.

About 90% of songbirds are monogamous, staying with one mate, but some species are polygamous. The male may mate with two or more females, a condition called polygyny; or the female may mate with two or more males, which is called polyandry. Researchers have found that some of the loveliest songs of the dawn chorus are produced by male birds practising this art of deception.

The song of the great reed warbler, for example – a migrant species with a relatively short breeding season – is an unusually elaborate and variable signal that changes according to its function. When a male is unpaired and is advertising for a female, he sings long and elaborate songs in order to attract her; once he has secured a female and paired with her, he ceases to produce love songs and defends his territory with a short, economical territorial song. However, when a polygyanous male wishes to attract a second female, he moves to the far end of his territory – as far as possible from his original mate's nest – and produces his long, complex song once more, so that any unsuspecting female passing that way will believe him to be unpaired and in possession of a good nesting ground. But the deception is double-edged: by the time the wayward male moves back to his first mate, a second male might well have moved in. Pied flycatchers practise similar deception techniques, as do house sparrows. The latter are not noted for their melodious song, but their monotonous under-the-eaves cheeping contributes to the overall effect of the dawn chorus. One British house sparrow colony, the subject of a lengthy study by scientists of the US Institute for Scientific Information in Philadelphia – completed in 1993 - was found to be rife with infidelity and incest. In one nest tests disclosed that the parents were mother and son, and in another nest of three young birds only two shared their genetic fingerprint with their `father', although all shared it with the mother. Blood samples from several male sparrows unmasked the adulterer, a bird from a nearby nest.

Why do air traffic controllers have to leave such a large gap between airliners landing and taking off on the same runway ❓

Even between two large passenger carrying aircraft, such as Boeing 747s, it would be normal to enforce a separation of at least 4 nautical miles. An executive jet would have to stay 8 nautical miles behind a 'Jumbo' in its final approach. This is obviously very inefficient in a busy international airport; indeed it has been estimated that fitting in one additional landing per hour at Frankfurt airport would save millions of dollars a year.

In that case, why is the distance between take-offs and landings so great?

Any type of wing will generate a vortex at its tip (See *Why does a sailplane have such long, tapering wings when compared with a powered aircraft?*) With large Jumbo jets, these can be extremely powerful horizontal tornadoes - one from each wingtip - several miles long and quite capable of overturning a light aircraft. Indeed, these vortices quite frequently rip tiles from roof tops near airports, especially during an aircraft's take-off when it is at its heaviest (because of the fuel load) and the wings are producing maximum lift. Indeed many hundreds of air accidents have been attributed to wake vortices.

Wing-tip vortices also generate a huge amount of drag which has to be overcome by engine power; this consequently adds to the fuel costs of the airline.

In the early seventies, researchers at the University of Oklahoma observed that hawks, eagles and buzzards gained their great ability to soar from the ragged tips to their wings which broke up the vortices at source, and they considered fitting slotted and flexible tips to aircraft wings. More recently, an elegant solution has been suggested and tested in a wind tunnel. This is to fit a small passive turbine to the wingtip.

The turbine is spun by the core of the vortex and absorbs much of its energy, replacing the twisting spiral of air with random turbulent flow. Drag on the aircraft is reduced and the spinning turbine can even be used to drive a small electrical generator.

!

Why does tonic water have a bluish tinge in certain lights (and what does it have in common with some washing powders) **?**

The distinctive taste of 'Indian tonic water' comes from small quantities of quinine sulphate. Quinine is an extract from the bark of the cinchona tree (usually Cinchona succirubra or red cinchona) which is native to the western mountains of South America but which was widely cultivated in the tropics after its medicinal properties became known in the mid-nineteenth century.

In the days of 'British India' quinine was certainly the safest and most effective treatment for malaria and has some beneficial effect in the treatment of other fevers as well. Westerners in malarial regions were in the habit of taking daily doses in the form of quinine sulphate to ward off the disease.

Tradition has it that they became so used to the taste and, indeed, actively grew to like it, that it was added to what we now know as tonic water. (An alternative tradition is that the taste was, by common consent, so obnoxious that it formed a socially-acceptable excuse to have a tot of gin with which to cover it up.) In fact, in any circumstances, there are sound reasons for drinking tonic water before a meal – whether mixed with the optional gin or not. Quinine sulphate induces reflex secretions from the salivary and gastric glands which strengthen appetite and speeds our digestion. (These secretions make you sick, however, if you take too much although it would be most unlikely that anyone would ever suffer this through drinking tonic water.)

Yes, but where does the blue come in?

The first person to record the bluish tinge in quinine sulphate was the English astronomer, Sir John Herschel, in 1845. This led to the discovery of what we now call fluorescence. (The phenomenon was at first thought by some to be a form of diffusion effect caused by suspended particles in the solution - see *Why is the sky blue?*) It was later found that if you move a test-tube containing dilute quinine sulphate through the bands of a natural spectrum formed from the Sun's rays, the solution remains transparent until it reaches the violet end, when a tinge of blue light is seen in the tube. If it is moved further into the invisible ultra-violet light, it becomes a vivid sky blue.

Strictly speaking, fluorescence is the emission from a

substance of any wavelength different from the one absorbed, but it is most commonly associated with visible emissions resulting from the absorption of ultra-violet light.

The ghostly blue in your tonic water is a result of this phenomenon, being an emission from the small amount – most of it is filtered out by the ozone layer – of normally invisible, ultra-violet light present in the Sun's rays. (If you think that tonic water is getting bluer than in days of yore, perhaps you should start worrying about the depletion of the stratospheric ozone layer.)

And washing powder?

Those washing powders which boast a result which is 'whiter than white' achieve this claim by including fluorescent substances in their composition – normally referred to as 'optical enhancers' (or somesuch) in the small print on the packet. (In some, more environmentally-aware, markets these additives are being phased-out in the new generation of 'green' washing powders since they are derived from petroleum and, as a consequence, are hopelessly non-biodegradable.) These impart a slight fluorescent blueness to the white in natural light. Some spectacular effects can be obtained in the theatre by blacking-out the stage and bathing it in invisible ultra-violet light. Any white clothes washed in 'optically enhancing' washing powder – or, indeed, any glasses of tonic water lying around the set – are the only things visible to the audience.

!

Why do we add salt to vegetables when we cook them ?

You might think that the reason we do this is that it is an easy and efficient way to make the vegetables taste salty. In fact very little salt will enter the vegetables and the best point at which to add salt to your personal taste is at the table. The reason that we add salt is to keep the flavour of the vegetables in.

Solutions have a tendency to become equal in concentration. Dissolved substances will tend (if they can) to flow from a stronger to a weaker solution while water will tend to flow in the reverse direction. We can see an example of this when we soak prunes in water overnight before cooking them. Attracted by the very high concentration of sugar within the dried fruit, the water flows through the skin of the prune and rehydrates it – which in this case is desirable.

A similar situation would apply in a pot of simmering vegetables if we didn't add salt to it. The water would flow into the vegetables, making them excessively 'mushy' while the dissolved substances within the vegetables which give them their flavour would seek out the pure water in the cooking pot and dissolve in it. At the end of cooking, the water you drained from the vegetables would probably have as much flavour and nutritional value as the vegetables themselves.

It doesn't much matter what is in the solutions – it is the relative concentration (measured by the mass of molecules of a given substance in a given volume of the solution) which matters - and you could achieve the same effect using sugar. However, salt is usually cheaper (and less sticky), and by adding it to the cooking water we can provide a sufficiently concentrated solution to prevent the 'leaching away' of flavours and nutrients.

Fringe benefits on the 'salary' (our word comes from the Latin salarium, *the money paid to the Roman soldier to buy his salt.)*

When an impurity is added to a liquid - in this case salt to water - one of the results is to raise the boiling point of that liquid. (See *Why is the red 'danger zone' on a car's temperature gauge normally around 120 degrees Celsius rather than at 'boiling point'?*) This means that by adding salt you can 'push' the boiling point, and hence the cooking temperature, to several degrees Celsius above 'normal boiling point'. Apart from being a slight boon to the impatient cook, the consequent reduction in cooking time is another factor in helping to preserve the nutrients in the vegetables.

!

Why throughout the world does the Captain of an airliner always sit in the left hand of the two pilots' seats ▮

In the air, as on the roads and on the high seas, there are rules, which, if observed by all, help to avoid collisions.

In the early days of flying the basic navigational instruments of pilots were a map, a compass, a watch and the human eye. The easiest means of finding the way was to follow roads, railway lines, coasts and any other linear geographical feature. It seems that while concentrating on these line features, their maps, flying the plane and generally keeping an eye on things in the cockpit, pilots sometimes met head on when they were flying in opposite directions but using the same feature to navigate by.

To avoid this, it was agreed that all pilots should always keep a line feature to their left when following it and thus avoid such collisions. As aircraft grew bigger and wider, a seat in the middle of the fuselage became increasingly inconvenient as vision was restricted, especially on landing and take-off in aircraft with a very 'nose-up' attitude. As the range (and also the time spent in the air) of aircraft became longer, it became usual to have a second seat in the cockpit alongside the Captain for a navigator or second pilot to assist the Captain. Naturally, the more senior of the two officers took the position on the left hand side.

This is more than just a traditional convention, however. To this day, even with electronic navigational aids, it is

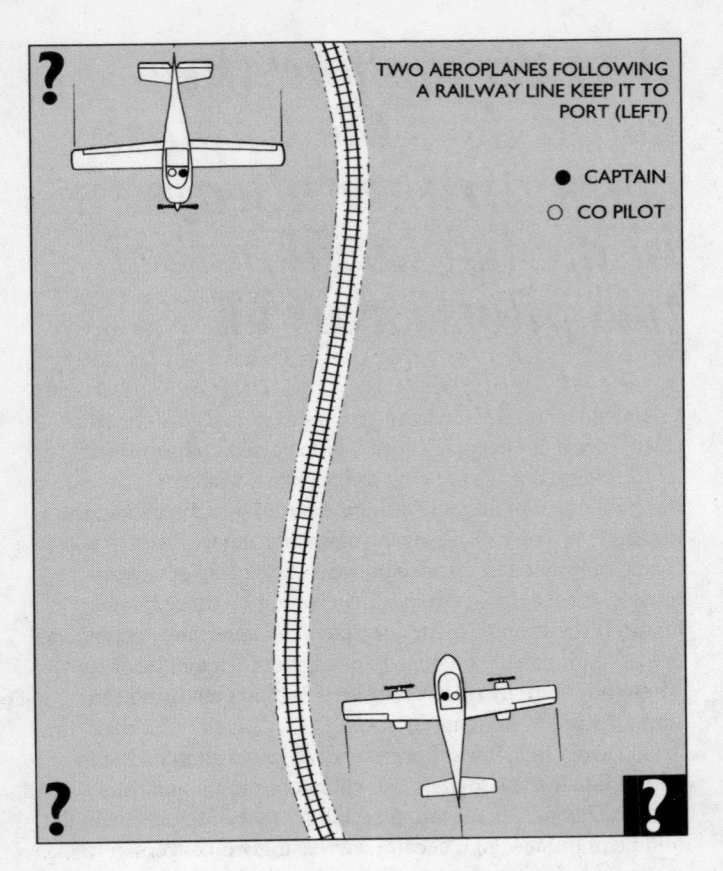

TWO AEROPLANES FOLLOWING
A RAILWAY LINE KEEP IT TO
PORT (LEFT)

● CAPTAIN
○ CO PILOT

reassuring to know that if all else fails, the rule is adhered to
and that for instance two pilots following the same railway line
in opposite directions are less likely to meet head on.

Why do clocks run clockwise ❓

In some senses this may seem a totally daft question – by definition, in any linguistically logical universe, clocks could only run 'clockwise' whether this meant from bottom to top along a straight line and back again, or whatever. However, what we mean here by 'clockwise' is the generally accepted sense of a hand sweeping from left to right at the top half of its cycle and from right to left at the bottom.

There is nothing mechanically 'natural' or determined about the direction in which a clock runs, in that it is no easier to build a clock which turns in one direction rather than another. Yet why do clocks – apart from 'novelty' items sometimes used to confuse the patrons of public houses – generally turn the same way?

It seems reasonable to assume that as the first mechanical clocks with dials were developed in the Northern Hemisphere it was more natural that the pointer or hands moved in the same way as the shadow of a sundial. While the Sun rises in the East in the Southern Hemisphere, just as it does in the Northern, the shadow of a sundial moves in the opposite – or anti-clockwise – direction.

This accident of history runs rather deep in our mechanical culture in that almost any form of dial giving a read-out of quantitative information – from an ammeter to a rev. counter – will indicate an increase in the quantity being measured by moving clockwise. Perhaps the increasing use of digital read-outs for just about everything – including time – will free us from this unconscious 'hemisphere-ism'.

Why can't animals talk ?

It is a well-established fact that all animals communicate with one another in some way. Many can communicate with human beings through the use of sign and body language to express their needs. But only human beings possess the biological machinery that makes speech possible, enabling them to use words as a means of communication.

Many scientists believe that the evolution of speech was as crucial to man's mastery of the world as the evolution of his brain. After all, some animals display an awareness of the world about them, in the same way as humans, although they may express intelligence in a different way, so the argument is that the power of speech gave us the boost that was necessary to turn us into the dominant species.

Human speech depends on two factors: the brain and the vocal tract. Apes are very intelligent primates, possessing an advanced communication system, but they do not have the human ability to produce speech. Why is this?

According to generally accepted scientific opinion, the hominid and ape lines began to diverge about 8 million years ago, and the predecessors of Homo sapiens emerged some 4 million years later. At some point after that, the organs that produce sounds in all animals adapted subtly in man's ancestors to the point where they were able to produce rapid, articulate speech. No one can tell how, or over what period of time, this adaptation took place, because no fossil remains of speech organs - which are composed of relatively soft tissues that decay soon after death - have survived.

In all mammals, sound begins at the vocal cords, which are located in the larynx - the structure in the neck that creates the bulge known as the Adam's apple. The sound is then modified by the supralaryngeal vocal tract, which consists of three separate resonant chambers - the mouth, the nose and the pharynx, the area at the rear of the throat where the nose and mouth meet the oesophagus. Complex neural mechanisms control the distribution of sound within these chambers by co-ordinating the interplay of the soft palate - the continuation of the roof of the mouth - the tongue and the lips.

The vocal cords themselves act rather like a pair of lips, constricting the larynx. They open and close repeatedly in response to air being pushed up from the lungs, releasing small puffs of it to the speech-producing organs. The vibratory frequency of the vocal cords depends on how much tension is imposed on them, and we have control over that; we can choose whether our voice is to be high- or low-pitched.

Other primates, such as chimpanzees, have vocal cords very similar to ours, and the sounds they produce are also similar; yet they are unable to turn those sounds into speech.

The reason seems to lie not in the structure of the sound-producing organs, but in their location. The human larynx is lower than a primate's, making the pharynx larger, and the tongue - which lies entirely within the mouth in the case of a primate - extends down into the throat. Scientists have compared a primate's vocal tract to a bugle, which has a fixed length of tube and cannot be modified. A human's vocal tract, on the other hand, resembles a trumpet, which is more versatile because it has three valves that modify sound by directing it down several passageways. This allows sound coming from the mouthpiece to be modified, something that is not possible with a bugle, even though the instruments are made from the same materials and make more or less the same initial sounds.

Human babies are born with the so-called 'standard' vocal plan similar to that of a chimpanzee, but shortly after birth the larynx begins to descend and the distance between the roof of the mouth and the rear of the base of the skull decreases, making more room for speech modulation. By the age of six a human child is capable of making all the sounds necessary for fluent and articulate speech. This does not happen with any other mammal, which is why speech is unique to human beings. However, there is a disadvantage. Since the opening to the trachea, or windpipe, is situated low in a human being's neck, we have to be careful how we swallow our food in case it 'goes down the wrong way'. Other mammals do not have this problem. As their windpipe makes direct contact with the nasal passage, the food they swallow passes to the left and right of the larynx. Most can breathe through their noses while drinking or swallowing - something we cannot do.

!

Why, if you use a laptop computer, should you vary your daily workload **?**

There are two commonly used types of rechargeable battery - lead-acid as generally used in a motor car and nickel-cadmium or 'NiCad'. Because of their better power to weight ratio, NiCad batteries, although initially more expensive, are generally preferred in applications such as laptop computers, cordless drills, portable telephones and electric shavers where lightness and portability are of the essence. NiCad batteries also have the advantages that they can usually, if properly looked after, withstand more cycles of charge and discharge (which offsets their higher initial cost) and have a more consistent power delivery through the life of each charge, although their performance drops off very rapidly at the end of a charge. Characteristically, the performance of a lead-acid battery gradually, but continuously, deteriorates as it discharges.

NiCad batteries do, however, have a quirk which it is worth being aware of. If they are always discharged by the same amount and then fully recharged before the next use, they develop a 'memory' of the level of charge to which they are habitually run. Suppose, for example, you always work for an hour in the evening on your laptop machine before putting it back on overnight charge. Each work session will run down about a third of the life of a full charge. In the course of time the battery becomes accustomed to working at this level and becomes very reluctant to give up the 'forgotten' two thirds of its remaining charge. This, of course, presents a problem when you break your habit and decide to do a solid three-hour stint on the machine or for some reason are unable to recharge it between work sessions - you have 'educated' it to run at a third of its capability.

For this reason, the manufacturers of all appliances which use NiCad batteries recommend that, every three months or so, you 'exercise' the batteries by discharging them completelely and leaving them discharged for about twenty hours before recharging them in the normal way - as you will have seen if you ever read the 'unexciting' bits of the technical documentation which tend to be skipped on Christmas Day.

(Incidentally, lead-acid batteries should never be discharged in this way as irreversible chemical changes take place below a certain level of charge and the battery becomes useless for anything other than scrap.)

!

Why does the salting of road surfaces in winter make steel-bodied cars rust more quickly ❓

Lay not up for yourselves treasures upon earth, where moth and rust doth corrupt, and where thieves break through and steal.
St Matthew, Chapter 6, Verse 19

After death and taxes, rusting or corrosion is probably the third great certainty of modern life. For example, from the moment a car is built most of the metals of its construction are striving to achieve a more stable chemical state - often akin to the ore from which they were extracted in the first place. Rust has always been a source of trouble for life on earth. Over two billion years ago when the first organisms were generating oxygen by photosynthesis, the oxygen was not free to accumulate in the atmosphere or in water. Instead, it was being grabbed by iron in the formation of rust.

Rust is hydrated iron oxide and in order for it to form, both water and oxygen must be present. (If you put an iron nail in a sealed bottle of water from which all of the dissolved oxygen has been driven out by boiling, it will not rust.) Unfortunately (in this context), both oxygen and - unless you live in Arizona - water are available in abundance in the environments in which most cars operate for at least a part of the year.

Rusting is an electrochemical reaction in which the iron molecules lose electrons which enables them to combine with oxygen. In other words, there is a minute electric current flowing away from the iron. Very pure water is a surprisingly poor conductor of electricity. However, in most everyday occurrences of water there are enough impurities in it to make it a very good conductor indeed and when road salt (sodium chloride) is added to the mix, it acts as an electrolytic conductor and speeds up the reaction by increasing the electron flow from the iron - or more normally steel. (Steel is an alloy of iron and carbon, of which more anon.) In such technologically spendthrift societies as the UK and USA this chemical barbarism of spreading salt on icy roads is commonplace. In more enlightened parts of the world such as Germany, road salting is strictly forbidden, although the reason is as much to do with avoiding the increased salination of streams and rivers.

However, the story of corrosion and the attempts to combat it does not stop here. In any machine or engineering structure and even in the kitchen cupboard you can see evidence of the thought and hard-earned experience which has gone into controlling corrosion.

Why do metals react with each other?

Metals fall into what is known as a 'reactivity series' which is arranged (in descending order) by their reactivity or, in other words, propensity to give up electrons. The order of some of the metals frequently used in engineering is given by the mnemonic 'MAZIT' for -

> **Magnesium**
> **Aluminium**
> **Zinc**
> **Iron**
> **Tin**

Of the other frequently encountered metals, lead and copper (in that order) lie below the above series. At the very bottom of the series are gold and platinum (again in that order) which to most intents and purposes do not react at all.

This is why they are usually found in uncombined forms in nature and do not have to be extracted from an ore.

When a metal high in the series is in contact with one lower down, an electrolytic cell will tend to form. Basically, electrons will flow from the more reactive to the less reactive metal and given the presence of water and oxygen the corrosion of the more reactive of the two metals will be accelerated. For this reason an engineer has to be constantly mindful of what metals are in contact with each other in any structure or machine. For example, it would be unwise, indeed bordering on the insane if you value your life, to attach a lead keel to the hull of a boat using steel bolts. Similarly, screwing a copper name plate to a door with steel screws will produce an unsatisfactory result. In both instances the electron flow from the steel will accelerate its corrosion. This phenomenon is referred to as 'electrolytic' or 'galvanic' corrosion.

Putting this to good use

However, like many things in nature, once these effects are understood they can be turned to our advantage. For example, corrugated iron sheeting is usually coated with a layer of zinc - a process known as galvanizing. Because zinc is higher in the reactivity series the electron flow is from the zinc to the iron. Even if the coating is imperfect or becomes damaged, the zinc 'sacrifices' electrons to the iron and continues to protect it. The reason the zinc itself does not rapidly corrode away to nothing is that the product of the corrosion itself, zinc oxide, forms a reasonably impermeable surface skin which, provided it is not disturbed by abrasion, protects the underlying metal from contact with further air and water. Galvanizing is not always a convenient process, but many anti-rust primer paints of the sort you would buy in a car accessory store or ships' chandlers are rich in zinc salts. If used correctly on steel, the principle is exactly the same.

On a larger scale, 'sacrificial' plates are frequently attached to the hulls of metal boats to protect the hull. This was first proposed by the English chemist, Sir Humphry Davy in 1823. He suggested to the British Admiralty that the copper-bottomed hulls of their ships could be protected by fixing zinc blocks to them. The principle was absolutely sound, but the scheme was not adopted as it was found that the corrosion-free bottoms were more attractive to barnacles which generated undesirable drag on the hull. (Until recently, when worries about the pollution of harbours came to the fore, copper salts were a constituent part of many marine anti-fouling paints.) Davy was so angered by the Admiralty's failure to develop his suggestion that the incident contributed to his decision to retire from public life. Nowadays, it would be normal to use zinc blocks in contact with a steel hull. The zinc corrodes freely in the electrolytic salt water and so has to be repeatedly replaced. However, the purpose is achieved as it is cheaper to sacrifice a few blocks of zinc than to pay for a new hull.

Food 'tins'

The reverse effect can be seen in food 'tins' which are actually made of mild steel coated with an alloy of tin and antimony or bismuth which prevent the tin from going powdery at low temperatures. Tin being fairly low in the reactivity series is a useful coating in that it does not react with acidic foodstuffs such as fruit. However, the moment the surface is scratched through to the underlying steel, electrolytic corrosion sets in and the steel rusts very rapidly – as can be seen on any waste tip.

... and back to car bodies (and sword blades via blacksmiths)

As we said earlier, steel is an alloy of iron and carbon. The carbon is added to give the material hardness and strength – metals tend to be relatively soft and weak in their pure state – but this addition actually increases its likelihood of corroding.

Just as there is an electron flow from zinc to iron in corrugated sheet between the two metals, so there is one from the iron to the carbon *within* the alloy, but in this instance it is wholly detrimental to the anti-corrosive properties of the material.

So why do blacksmiths hit iron with hammers?

This can be seen by comparing steel with old-fashioned wrought iron which has had all of the carbon laboriously hammered out of it by the blacksmith. (If you are wondering why there is carbon in the iron in the first place, it got there from the charcoal or coke which was used in extracting the iron from the ore.) We cannot go into the mystery of blacksmithing here – it is one of the most fascinating subjects on the face of this earth – but in making a very high quality and important item such as a sword blade, the smith would heat the iron till it was cherry-red, hammer it flat, fold it over, hammer it flat again and repeat the process many hundreds of times. This meant that impurities were constantly driven to the surface and, at this temperature, the carbon would react with the oxygen in the atmosphere and escape as carbon monoxide gas. There is many a mediaeval sword which is still around after a bloody life in battle followed by centuries of neglect, while breakers' yards are full of twenty – year old cars which have been scrapped because of structural corrosion.

If you are beginning to wonder why car bodies are ever made of steel - it is a very good question. However, it is one you should ask a politician rather than us. Take him, preferably in a 1960s Chevrolet Corvette or Lotus Elite, to the depressing sight of a large commercial breakers' yard when you ask it. Apart from those that have been crashed, the great majority of vehicles are there because of structural corrosion - not because their mechanical components have worn out.

!

Why do we add ice cubes to our drinks (and when does ice sometimes warm a drink) ?

The obvious answer to the first question is to keep the drink cool, but the precise way in which the ice cubes do this should not be regarded as altogether straightforward. For example, a similar volume of water at 0 degrees Celsius would not keep a drink as cool for as long.

When we say that something is cold, all we mean is that it has less energy than something we would regard as relatively warm. By definition, and the second law of thermodynamics, a lack of energy cannot 'do' anything when confronted with an excess of energy. While it is perfectly reasonable to draw a curtain to 'keep the heat in', it is totally illogical - or, at best, rather unscientific – to do the same thing to 'keep the cold out'.

What the ice is doing is to take up heat energy from the drink, but not just because the ice is cold. The heat energy is, in fact, being used to convert the ice from the solid to the liquid state. In much the same way that a pan of boiling water cannot exceed 100 degrees Celsius (unless it is pressurised), the ice cubes cannot rise above 0 degrees Celsius. In both instances all of the energy goes into the change of state without causing any change in temperature.

Indeed, if the drink is an alcoholic one, the ice cubes will be even more effective in cooling it – for a while. Although the ice will melt faster the alcohol lowers the melting point and the ice takes up more heat while it melts. (This phenomenon can cause a great deal of annoyance on a frosty morning. If you use a de-icer spray on the outside of your car's windscreen, the antifreeze in the spray melts the ice by bringing the melting point, and hence the temperature of the glass, rapidly down - usually to about -15 degrees Celsius. You will then sit for a further five minutes waiting for your breath to unfreeze from the inside of the screen.)

As a final aside on the subject of ice cubes, we often find it deeply amusing when people make a fuss about ordering a particular brand of pure spring water and then proceed to adulterate it with ice made from the humble tap variety.

...and ice warming drinks?

One of us noticed this phenomenon on a wickedly hot and humid summer's evening on the shores of the Baltic in northern Poland. The waiters were serving tiny glasses of neat vodka - glass and all - from a deep freeze, possibly at a temperature as low as -20 degrees Celsius. A film of ice very quickly formed over the outside of the glass. In order to first condense and then form ice, the water vapour was obviously having to give up heat energy to somewhere – in this case the glass and its contents. The ice was warming the drinks!

!

Why did Formula One racing teams once chill their petrol ❓

In the 'turbo era' of Formula One racing in the early 1980s the tendency of most teams was to produce engines which were hugely powerful but very inefficient in terms of fuel consumption. For the 1984 season, in an attempt to force the cars to be more fuel-efficient, a rule was introduced limiting the petrol tank to a capacity of 220 litres.

Since the essence of motor racing is to look for loopholes in the regulations in order to gain a competitive advantage, the engineers turned their minds to ways round this rule. If you chill petrol its density increases by roughly 1% for every degree Celsius drop in temperature. This meant that at around minus 50 degrees, what was 235 litres at normal temperature and pressure could be 'shrunk' to the statutory 220 litres and this stratagem was adopted by all of the major teams. However, they had a tense moment at the opening race of the season in Brazil. The start of the race was delayed by half an hour and there was the obvious worry that the tanks would explode as the fuel in them expanded under the fierce Brazilian sunshine. In fact they probably need not have worried *too* much as petrol takes up heat and consequently expands very slowly and indeed, the race started without any explosive incidents.

It is worth remembering, perhaps, that you are getting slightly better value for money if you buy it, since petrol is metred by volume rather than by weight, in the chill of the morning rather than in the heat of the midday sun.

Why does grass look white by moonlight ?

Although this phenomenon is becoming more difficult to observe because of light pollution, it can be very marked in rural areas, remote from street-lights, under a full moon, when even whitewashed cottages can seem unexpectedly dull in comparison with the grass.

The eye responds to different intensities of light in different ways. At the back of the eye there are two different kinds of nerve endings - cones, which are sensitive to bright light (especially yellow) and rods, which are sensitive to very much lower intensities of light. However, the rods behave rather like black-and-white film and are only able to react to light in a colour-blind manner. (This is nothing to do with the, usually inherited, red-green colour-blindness which is caused by a lack of the particular nerve-endings which respond to red light.)

The rods are receptive only to blue and green and, consequently, because of the monochromatic nature of their reaction, the reflected light from green vegetation appears brilliant white.

Do carrots help you see in the dark?

Because the glare would otherwise be excessive, the rods 'switch themselves off' under exposure to bright light. This is because rhodopsin, the light-sensitive chemical in the rods, is rendered inactive by bright light. The time taken for the body to synthesize fresh rhodopsin is the cause of the delay in the return of night vision after (say) being dazzled by car headlights. This process is dependent on the body's vitamin A. As carrots are a good source of vitamin A, we can trace the basis of the, unspeakably tedious (and generally unwarranted), exhortation familiar to millions of children in the western world to 'eat up your carrots', because 'They'll help you see in the dark'.

Why is a twelve-bore (gauge) shotgun bigger than a twenty-bore ?

It might seem natural to expect the firearm denoted by the larger number to have the barrel with the larger bore or gauge or calibre (all terms for the internal diameter), but the reverse is, in fact, the case. Why?

This method of measuring firearms goes back to at least the early sixteenth century when the Spanish matchlock was standardized as having a bore which fitted a ball of lead weighing exactly one tenth of a pound – an early example of decimalization without metrication, perhaps. This weapon rapidly became the standard infantry musket throughout Europe and with it spread this way of measuring the barrel of a gun.

Consequently, a modern twelve-bore shotgun has a barrel with an internal diameter equivalent to the diameter of a ball of lead weighing one twelfth of a pound, a twenty bore one twentieth and so on. Obviously the ball weighing one twelfth is larger than the one weighing one twentieth of a pound and hence the relative sizes.

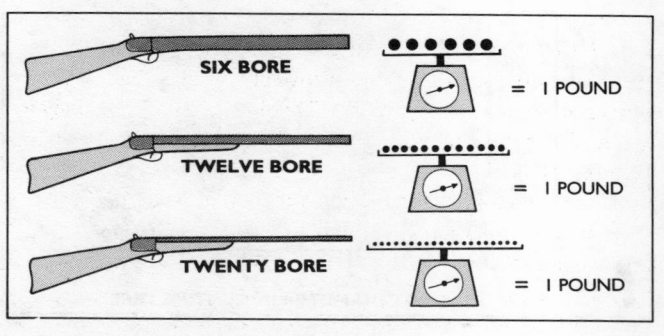

It is, however, a bit more interesting than that. Because the relationship between the volume (and hence the weight) of a sphere and its diameter is a non-linear one – ie a sphere with twice the volume does not have twice the diameter - the bore of guns is also non-linear.and a four-bore is not four times as big as a sixteen-bore.

As you will see from the graph a twelve-bore barrel is significantly larger in diameter than half the diameter of a six-bore.

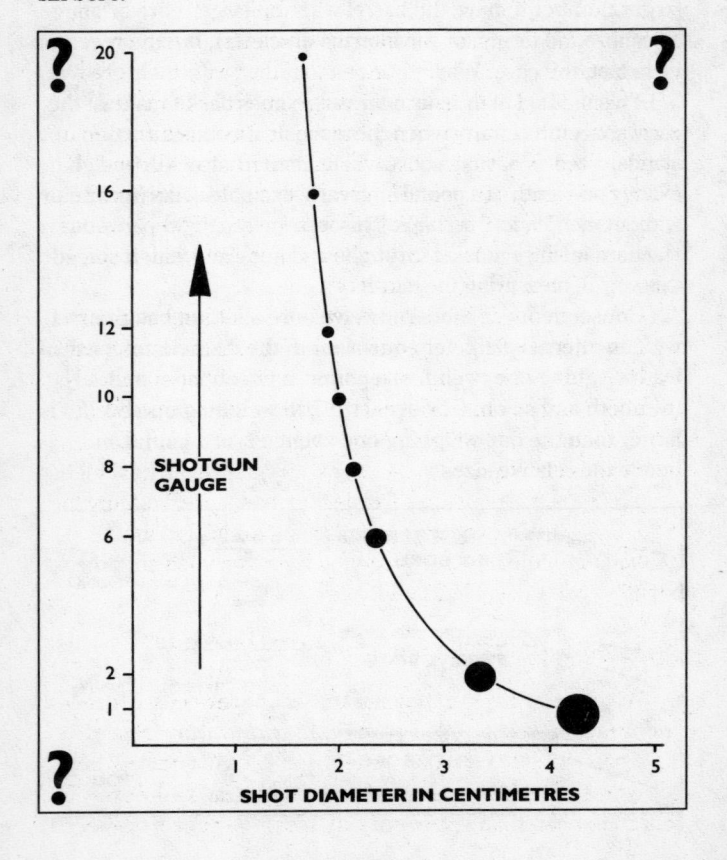

SHOTGUN GAUGE

SHOT DIAMETER IN CENTIMETRES

Why do woodworm ravage early versions of plywood but strictly avoid more recent types ❓

Early attempts to make plywood, in which thin layers or 'laminates' of softwood are bonded together with their grain at right angles to that of their neighbours in order to make a board with no overall grain and hence no obvious direction in which to split, were fairly disastrous in their strength and ability to withstand damp. They were also particularly prone to infestation by woodworm. Yet modern (post-1960s) plywoods can be superb structural materials and suffer from none of these problems. Why?

Until the advent of modern synthetic resins, most glues were based on animal products. One of the then commonest wood glues, casein, was derived from milk. Other commonly used glues were made by rendering down the bones and hooves of slaughtered cattle with water. The structural problems with these early plywoods was that the animal glues were hygroscopic - in other words they had a great affinity for water and tended to absorb it very readily. This led to a weakening of the bond between the layers of wood and to buckling and infestation by rot and moulds.

But what about the woodworm?

Woodworm are the larvae of the furniture beetle (*Anobium punctatum*) and can cause great destruction to any form of dead timber such as beams, rafters and even furniture. They do not eat your furniture out of pure wickedness, however. In order to generate the strength with which to pupate they are struggling to build up nutrients. Among the most important groups of these nutrients are the protein-building nitrogen compounds. Most woods are very low in these, so the grub has to burrow through a huge amount of wood (relative to its own mass) in order to acquire an adequate sufficiency. In general they will consume any form of wood, whether hard or soft, and one has to assume that the greater energy cost in burrowing through a hardwood is offset by the greater density of nutrients contained within it. If this were not the case, woodworm would have evolved to specialise in particular types of wood because of the particular advantages they conveyed. In fact, woodworm are in this respect true generalists. By introducing animal-glue plywood, however, man was, as ever, interfering with the balance of nature since the cattle-derived glues were very rich in ready-made protein. No wonder this was attractive to the woodworm. It must have been the grub's equivalent of eating a steak sandwich.

There is no such thing as a free lunch, unfortunately, and this situation did not last. With the increasing use and sophistication of synthetic glues from the 1960s onwards, good-quality plywood became an excellent water-resistant material - and at the same time the woodworm lost their taste for it as the gravy train of effortless protein became a thing of the past.

!

Why should a kettle be shiny or white (and sometimes black) ❓

Shiny surfaces are very poor radiators of heat (see *Why does aluminium foil keep things hot?*). It therefore obviously makes sense for an electric kettle, in order to minimize heat loss through radiation, to be either chromium plated or made of highly polished stainless steel.

However, metals are good conductors of heat and for this reason electric kettles have been increasingly made of white or lightly coloured plastic. This combines the benefits of cutting down heat loss by conduction (and hence ultimately by convection) and by radiation (since light colours radiate less heat than dark ones).

... and black?

However, if you have seen an old-fashioned arrangement, whereby a kettle hangs on a hook next to an open fire in order to heat it externally (as opposed to the internal heating of an electric element), the best colour for the kettle is matt black. As well as being the best radiator of heat, black is also the best absorber. The dirtier and smokier the outside of the kettle, the more efficient it will be and to polish it would be a false move in 'home economics' terms.

When it comes to the teapot, however, if it is a silver one, it makes very good sense indeed to keep it highly polished – and then cover it with a white knitted tea-cosy.

❗

Why does bacon 'spit' when you fry it ❓

Commercially cured bacon contains surprisingly large quantities of water - as much as 20% by weight - when you take it off the supermarket shelf and pay for it (which may lead you to wonder exactly what is meant by a 'pound of flesh'). No matter how slowly you try to fry the bacon, there will come a point when the fat or cooking oil (which does not boil at 100 degrees Celsius like water and therefore continues to rise in temperature) will be considerably hotter than the normal boiling point of water.

If the water in the bacon were free to boil away slowly and progressively, this would probably not present any problems. However, this is not the case, since, by the time it is in the frying pan, the water in the bacon is trapped under a layer of hot cooking oil. In the laboratory, under carefully controlled conditions, it is possible to raise the temperature of small drops of water, suspended in a suitably dense oil, to as much as 180 degrees Celsius before they boil. In a rather less controlled manner, this is what is happening in the frying pan and when the drops of water eventually do boil, they will do it explosively, expanding to about 1800 times their original volume in a fraction of a second, projecting droplets of very hot oil indeed around the kitchen.

Apart from making a mess, this can be quite dangerous and it is wise to be very wary indeed of plunging any food with a high water content (or ice content) into a pan of hot oil or fat. Equally, you should always try to remember, if confronted by a chip-pan fire, never to attempt to plunge it into a sink of water or attempt to douse it by using water. The 'flash' boiling of the water is liable to cause instant and very serious scalding.

Why is a 'state of the art' wind tunnel refrigerated ?

While modern computerized modelling systems can be very sophisticated indeed, the black art of aerodynamics is so complex and critical that most aeroplane manufacturers still prefer to put their faith in the traditional way of testing a new design – ie. in a wind tunnel. Because it is very expensive in terms of materials and resources to use a full-size mock-up of a new aircraft – remember it might have to be scrapped and rebuilt many times in the course of researching all of the different variables – it is normal to use scaled-down models. However, this presents a problem. While you can scale-down the aircraft you can't scale-down such an amorphous thing as the medium in which it operates air. Or can you?

In hot air, the molecules are less closely spaced than in cold air (See *Why does a hot air balloon rise?*). Equally, by chilling the air you can make it thicker and, although you can't reduce the size of the molecules themselves, you can in effect 'shrink' the air by reducing the spaces between the molecules in proportion to the scale of the model being tested. Indeed, this is exactly what is done at the European Transonic Windtunnel near Cologne in Germany. This has the capability of running at temperatures as low as -180 degrees Celsius. At this temperature pure nitrogen has to be used in the tunnel because the oxygen in normal air would liquefy.

Although it is a very expensive facility to operate, manufacturers of aircraft regard the cost to be worthwhile because of the potential fuel efficiencies to be gained. We do not, however, regard it as essential that, in the pursuit of complete authenticity, model train sets have to be run in 'scaled-down' air.

!

Why do some car windscreens appear mottled in the evening sunlight ❓

(For reasons explained below, inhabitants of some countries, in particular the United States of America, may never have seen this phenomenon.)

For centuries, glassblowers have known that if you drop a dollop of molten glass into a tub of cold water, the result is an extremely tough, teardrop-shaped solid which can be dropped on a stone floor or robustly hammered without its shattering. However, if pressure is applied to the pointed end of the tear-drop, the whole tear-drop shatters into a pile of fairly uniform, roughly cuboid, small granules.

PHASE 1 PHASE 2 PHASE 3

LIQUID GLASS

PLOP

LIQUID

SOLID

WEAK TAIL

STRONG SKIN HELD IN COMPRESSION BY SHRINKAGE IN CENTRE

(These teardrop-shaped curiosities are known as Prince Rupert's drops after the nephew of Charles I of England who, apart from being one of his uncle's leading commanders in the English Civil War, was also an enthusiastic amateur experimenter and materials engineer. Prince's metal, an alloy similar to brass, is also named after him.)

The reason for their behaviour is that when, during their formation, their surface is suddenly cooled it becomes rigid. However, because glass is a rather slow conductor of heat - which is part of the reason ordinary glass jugs shatter when you pour boiling water into them – the interior of the tear-drop is still soft and still trying to 'catch up' in its contraction as it solidifies. This means that the surface of the eventual solid tear-drop is under considerable compression - is being 'pulled in', if you like, by the core of the tear-drop which is in tension. As a result, the surface is fairly immune to further external compression, such as the blows from a hammer.

The weak point, however, is the pointed tail of the tear-drop where, because there is so little material, it is fairly easy to get at the considerable stresses in tension inside the tear-drop and, by applying pressure, cause it to shatter.

This laboratory plaything is the basis of all toughened glass as used in applications such as car windscreens where it would be undesirable to have dagger-like shards of plate glass flying around after its tensile failure in an accident. In toughened glass the compressed surface helps to restrain these dangerous projectiles.

The mottles

In the process of toughening, a car windscreen is rapidly cooled from the near-molten state using jets of cold air. The mottled effect you can sometimes see is the 'ghost' of where these jets of air hit the surface. The reason this effect is most frequently seen in the evening, is that there is a greater proportion of polarized light at this time of day. Polarized light (which has a simpler wave form than normal light) refracts in glass in such a way as to show up the stress patterns. (Because of this the optical glasses used in astronomical telescopes have to be very evenly cooled over many hundreds of hours so that there are no internal stresses caused by uneven contraction.) These mottles can also be seen if you wear certain types of sun-glasses which have polarizing filters. These make driving a car with a toughened windscreen almost impossible.

Why only some car windscreens?

While toughened glass is obviously preferable to ordinary plate glass in that it breaks into small, relatively harmless chunks with concave edges (because at the moment of collapse the interior of the material achieves the dimensions it has been striving for since the moment of its formation), it does have some disadvantages. Not least of these is that it is

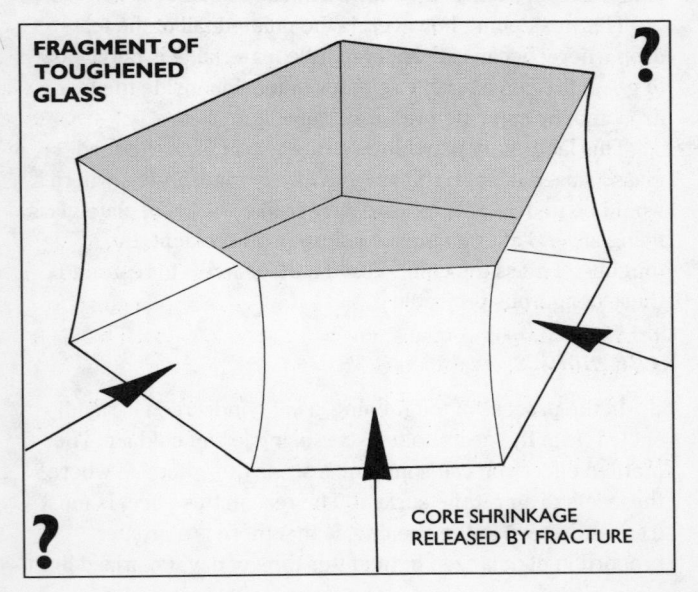

FRAGMENT OF TOUGHENED GLASS

CORE SHRINKAGE RELEASED BY FRACTURE

rather tougher than the human skull, which may, after all, come into violent contact with it. It also becomes impossible to see through when it does shatter – possibly when hit by a stone chipping – which can be alarming when driving at high speed on a crowded motorway. For these reasons it has been illegal to use toughened glass in windscreens in the United States for many years and laminated glass which has more 'give' in it is used instead. In laminated glass a layer of a transparent vinyl material is sandwiched between two layers of ordinary plate glass. If the glass is cracked or splintered, the bond with the sheet of vinyl is usually strong enough to arrest the spread of the crack. Even in complete failure of the screen the bond is resilient enough to prevent any splinters of glass becoming detached.

(Observant movie buffs will have noticed that in American cop movies bullets either 'carom off the windshield' - to quote the Clint Eastwood character in *Magnum Force* - or leave neat holes with short radiating cracks; while in the British equivalent - until recently, when laminated screens became more common in the UK - the screen shatters.)

If you are one of our older readers or have visited a motor museum, you may have seen examples of the early, pre-World War II attempts at laminated glass. These were fairly disastrous because celluloid was used in the sandwich between sheets of plate glass. Not only does this go yellow with prolonged exposure to sunlight, the bonding with the glass was never particularly successful, resulting in some terrifying-looking remains of windscreens to be seen lying around in scrap - yard wrecks.

Nowadays a very thin layer of a synthetic material such as polyvinyl butyral, which is incredibly tough and does not discolour with age, is generally used.

!

Why is mercury used in thermometers ?

Since, with very few exceptions such as water and antimony, most substances continuously (but not necessarily at a consistent rate) expand with an increase in temperature, it would seem reasonable to be able to make a thermometer out of pretty much anything. Indeed, at a pinch, you probably could and, with a great deal of trial and error over a wide range of temperatures by using an already existing thermometer as a reference point, calibrate it reasonably accurately. In fact. the first thermometer, or 'thermoscope', built by Galileo in 1596, used the expansion of air in a glass bulb to displace a column of coloured liquid. This was very sensitive but was unfortunately too greatly affected by changes in atmospheric pressure to be of any use as a measuring instrument. Galileo then proceeded to construct a thermometer using alcohol in a sealed glass tube very much like the ones which are familiar today.

However, a problem became apparent when it came to calibrating liquid thermometers of this type. Early attempts involved taking two fixed points and dividing the range in

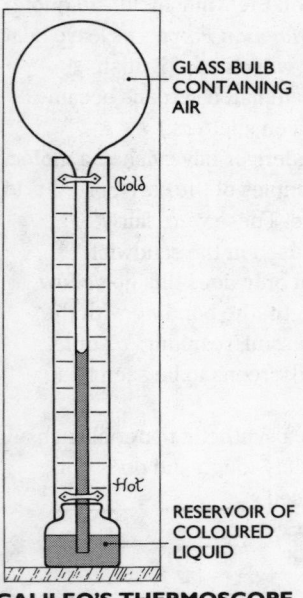

GLASS BULB
CONTAINING
AIR

Cold

Hot

RESERVOIR OF
COLOURED
LIQUID

GALILEO'S THERMOSCOPE

between them into equal increments. In the case of one of the earliest scales, proposed by the German physicist Fahrenheit in 1714, zero was fixed at the temperature of a freezing mixture of ice and salt (which is why the freezing point of pure water is at the seemingly arbitrary value of 32 degrees on the Fahrenheit scale), while the upper point was fixed at the temperature of the human body. Fahrenheit divided the range between the two fixed points into 12 equal increments or degrees. (It was only later that, in order to give finer gradations, these degrees were further divided by eight so that body temperature became 96 degrees on the Fahrenheit scale.) The problem which arose was that thermometers using different liquids gave different values at intermediate points between the two fixed ones on the scale. The only possible conclusion was that, quite apart from the fact that the liquids were expanding at different rates from each other, at least some of the liquids were themselves expanding at uneven rates across the temperature range, making it impossible to establish a unit which would be uniform at all parts of its scale.

Other scientists were working on the basis of a centigrade scale with the freezing point and boiling point of water being the two fixed points of respectively 0 and 100 degrees and Carlo Renaldini of the University of Pisa suggested that by mixing different proportions of water at 0 and 100 degrees you could create a mixture capable of establishing all of the intermediate values and thus calibrate a definitive thermometer. The theory was that equal weights of water at 0 and 100 degrees would produce a temperature of 50 degrees; a mixture of 80 per cent at 100 degrees and 20 per cent at zero would produce a temperature of 80 degrees and so on for all the intervals in between. Apart from being extremely difficult to carry out in practice, the flaw in this method is that it depends on a single substance, and a profoundly suspect one at that – water. Whether one measures the proportions of water by weight or by volume, the highly anomalous behaviour of the density of water at around 0 to 4 degrees renders it totally unsuitable for the establishing of a standard unit of temperature.

It was much later, in the early nineteenth century, following the work of Charles and Gay-Lussac, that it was shown that all gases expand by the same amount and at the same rate not only as each other but also across the entire temperature range. This led physicists to believe that there was something fundamental about the expansion of gases and indeed they were correct as was later proved theoretically by Lord Kelvin. In a sense, Galileo had, in his first thoughts, been right after all.

While, ideally, a gas thermometer will always be more accurate, in practice, a liquid thermometer is generally more convenient and of all known liquids mercury most closely approximates to the gas scale.

However, the story does not end there and there are other properties of mercury which make it uniquely suited to its use in thermometers. Mercury has a very low specific heat (the amount of heat energy required to raise its temperature) compared with other liquids and, being a metal, it is also very conductive of heat. This means that a mercury thermometer is very quick and reactive in response to small fluctuations in temperature. It has a freezing point of -39 degrees Celsius and does not boil till 357 degrees which gives the thermometer a very useful range of operation. It does not expand very much (about one seventh the expansion of alcohol) and this might be seen as a disadvantage but it can be easily overcome by making the bore of the thermometer very fine. However, the low expansibility of mercury reduces the need for a long thermometer stem exposed to the air. Such exposure can be a very serious source of error in other types of liquid thermometer when the temperature of the air is the last thing that you want to measure. Most miraculously of all, perhaps, is the fact that mercury does not wet the inside of the thermometer (See *Why is water wet?*), so that blobs of it don't stick in the bore and give a false reading.

The use of this metal does have a disadvantage, however. Mercury is very toxic (as is its vapour) and notoriously difficult to clean up. People have only relatively recently become fully alert to this and one dreads to think how many laboratories in the past were very unhealthy environments indeed because of small globules of mercury from shattered thermometers lurking between the floorboards.

!

Why is the sea salty

If you look at the mineral analysis printed on the label of even the purest, 'suitable for salt free diets', bottled spring water you will probably see something like this -

Bicarbonate	77 mg per litre
Calcium	22 mg per litre
Sulphate	12 mg per litre
Chloride	10 mg per litre
Sodium	6.8 mg per litre
Silicate	5.7 mg per litre
Magnesium	5 mg per litre
Nitrate	2 mg per litre
Potassium	0.3 mg per litre
Fluoride	0.07 mg per litre

The substances with the suffix '-ium' denote the positive ions of metals; those with the suffix '-ate' or '-ide' denote negative ions of non-metals. The presence of these ions is the result of the reaction of rainwater, which is slightly acidic, with rocks which contain these soluble minerals. (We worry, quite rightly, about the effects of 'acid rain' caused by pollution but it should be remembered that it is perfectly natural for rain to contain small amounts of carbonic, nitric and sulphurous acid - and always has been.) Ordinary table salt is sodium chloride - the combination of one positive sodium ion with one negative chloride ion.

The concentration of salts in sea water varies vastly in different seas world-wide and even at different depths within the same sea, but a typical analysis would go something like this –

Total percentage by mass of dissolved salts   4%
Of which –

Chloride	55%
Sodium	30%
Sulphate	7.5%
Magnesium	3.7%
Calcium	1.2%
Potassium	1%
Bicarbonate	0.3%
Bromide	0.2%
Others	1.1%

Look familiar? If you were to conclude that salt-water was, by and large, concentrated freshwater you would be absolutely correct. When the rivers have flowed down to the sea, the concentration of salts becomes greater in seawater because of the evaporation caused by the heat of the sun. The water driven off by evaporation leaves the salts behind and then falls again as rain. If the rain falls over land, the cycle starts once more with the rain leaching out more salts. This phenomenon is not restricted to the sea alone. Any body of water where the fluid loss through evaporation is greater than that through drainage by a river will become salty (provided that the evaporation is not offset by a net fluid gain through rainfall). The most famous example is the Dead Sea between Israel and Jordan which is roughly five times as salty as the oceans.

Similarly, the Sea of Galilee which, because of its low level, is not very effectively drained by the Jordan, is also slightly brackish.

However, the two mineral analyses above don't quite tally. There is, for example, a greater concentration of bicarbonates, calcium and sulphates in the spring water than in the sea water. The first two are accounted for by where this particular spring water came from - the Campsie Fells in Scotland. These hills contain a stratum of limestone (calcium carbonate) and indeed it is at this level that the natural spring water lies. Calcium carbonate is soluble in the carbonic acid of the rainwater, however carbonic acid is unstable and readily breaks down on exposure to the atmosphere into carbon dioxide and water. When this happens the calcium carbonate precipitates out of the solution as a solid. This is why stalactites form on the roofs of dripping caves in many limestone areas.

The lower level of sulphates in sea water is accounted for by the fact that they are taken up by algae to maintain their salt balance. (See *Why do we add salt to vegetables when we cook them?*) If the algae did not do this they would lose water from their (single) cells to the surrounding sea water and be unable to survive. In a complex series of reactions, the sulphur is returned to the environment when the algae die, in the form of dimethyl sulphide, which is the slightly pungent gas which gives sea air its distinctive 'bracing' smell. Although not fully understood, the breakdown of dimethyl sulphide in the atmosphere is known to be a critical factor in the formation of rain clouds – and hence the cycle continues…

It must be pointed out, however, that there are many complex and only partly quantified forces at work here. Recent theoretical calculations to predict how salty the sea 'ought' to be, have generally come up with answers which conflict with actual experience.

Why do trees shed their leaves ❓

Few of nature's miracles are as beautiful, or as complex, as the leaf of a tree. But it is far more than a merely aesthetically pleasing object. A masterpiece of natural engineering, evolved over millions of years, it manufactures food for the plant that bears it - and plants, in turn, ultimately nourish and sustain all land animals.

Leaves form an integral part of a plant's stem, possessing the same fibres and tissues as they develop in the bud. A typical leaf consists of a broad, flattened blade called the lamina, attached to the stem by the stalk. From the latter veins radiate outwards through the lamina; as well as transporting nourishing materials to and from the leaf tissue, they act as supports for the leaf, rather like the ribs in animals.

The process known as photosynthesis is the key to a leaf's food manufacturing ability. All plants contain chlorophyll, the green pigment which absorbs energy from sunlight, enabling the plant to build up carbohydrates from atmospheric carbon dioxide and water. All this chemistry goes on within the leaf's internal structure, which is protected by the leaf's skin: this is continuous with the stem's skin, so that there is no break that might allow the penetration of harmful agents from outside.

The central part of the leaf consists of soft-walled cells. About one fifth of it is made up of chloroplasts which contain the chlorophyll that absorbs the sunlight. The cells also produce enzymes - the proteins produced by all living cells - which act as catalysts in the chemical reactions on which life depends. In the case of the photosynthetic process, these enzymes act in conjunction with sunlight's radiant energy to break down water into its twin elements, hydrogen and oxygen.

The oxygen released by green leaves passes into the atmosphere through pores in the leaf's surface to replace the oxygen absorbed by plants and animals during the normal process of respiration, and also by other factors such as combustion. At the same time, the action of enzymes combines the hydrogen released from water with carbon dioxide to form the carbohydrates without which plant and animal life could not exist.

The chemical reactions that take place inside leaves, therefore, are vital to life on earth - which brings us to the original question of why nature permits them to drop off in the autumn. This process occurs every year in deciduous trees, and every two or three years in evergreens and conifers.

It is all a question of the leaf's primary role. Although it releases vital energy to sustain the world around it, its main duty is to sustain its parent plant, particularly during the latter's initial phase of growth, when it may not be able to derive sufficient nutrients from the soil through its root system. Although the process of producing sugar as a nutrient continues as long as the plant exists, it is never as vital as during the first season of growth.

The fall of leaves, whether completely in the autumn of each year in deciduous trees or at intervals of a few years in evergreens, happens because a weak area called the abscission layer forms at the base of the leaf-stalk or petiole. In the normal course of events, the fall occurs as days grow shorter and the process of photosynthesis slows down as less light reaches the leaves. When this happens, a band of soft cells forms across the base of the stalk and the leaf eventually drops. The wound is quickly closed by a healing scar that seals off the stem and prevents undue loss of moisture, which would be harmful during the inactive winter months.

Why should a fighter pilot not be too fit ❓

Because aeroplanes don't like going round corners on the level, a pilot has to either bank the aircraft into the turn or bank out of it. At the high speeds and rates of turn encountered in modern jet fighters, centrifugal force will, in the former course of action send blood to the pilot's feet and in the latter to his head. Since having a rush of blood to the brain is profoundly unpleasant - and the cause of, amongst other things, burst blood vessels in the eyes - most pilots will, if they have the option, bank into the turn. However, in extreme circumstances the flow of blood to the lower parts of the body can be so rapid that the brain is starved of oxygen and the pilot blacks out - undesirable in the middle of a life-or-death dogfight.

These forces are conventionally measured in terms of multiples of the force of gravity or 'G's. To put this in perspective, a good sports car, skilfully driven, can generate a lateral force when cornering of just over 1G (the Lotus Seven was the first road car to break this barrier in the 1950s) while a Formula One racing driver routinely experiences lateral forces of over 3Gs. This means that his head (plus helmet) is being pulled sideways with a force three time its own weight, which is why drivers of long standing have such visibly over-developed neck muscles and why novice drivers will sometimes wear a neck strap connected to the side of the car with which to brace themselves. In comparison, a fighter pilot may experience forces of over 7Gs, which if sustained for more than a few seconds will generally cause him to black out.

In order to counteract the effects of this pilots wear a 'G- suit' which has inflatable bladders across the stomach and thighs which help to restrict the centrifugal rush of blood to the feet. However, the state of the pilot's body itself also has a large part to play in resisting the effects of such forces.

Certain types of strenuous exercise, such as aerobics, have the effect of enlarging the heart, lowering blood pressure and decreasing the pulse-rate, which in normal circumstances is all well and good. However, when you need as much oxygenated blood as possible to be returned to the brain, the last thing you want is a large sump of a heart pumping slowly at reduced pressure.

!

Why, until the early 1960s did almost all cars in the world have a positive earth electrical system, but, almost without exception, a negative earth ever since ?

Corrosion occurs in a metal when electrons flow away from it (See *Why does the salting of road surfaces in winter make steel-bodied cars rust more quickly?*). This frequently happens in the obvious form of electrolytic action between two different metals in contact with each other, but it can have different, less visible causes as well.

To give an extreme example, a problem frequently encountered in boats occurs when the engine block is used as a convenient earth for the electrical system and is connected to the positive terminal of the battery. It is a somewhat crude, but generally useful, analogy to think of a battery as a sort of electron 'pump' - sucking in electrons at the positive terminal and pushing them out at the negative. In the example of the boat, this means that the battery is depriving the engine block of electrons and supplying them, via the battery's negative terminal, to the boat's electrical equipment. It is the potential difference between the earth and the negative circuit which allows useful work to be done.

Since the engine block is usually a fairly massive chunk of reasonably pure cast iron, which does not readily corrode and where a little surface corrosion would scarcely matter, there does not seem to be a problem. What people forget, however, is that their engine block is inextricably connected, physically and *electrically*, through the metal of the drive shaft, to the boat's propeller. Although less visible, it is just as destructive as if they had bolted bits of lead (as in the battery) straight onto their propeller. Since the propeller is probably made of steel and may even be an aluminium alloy it will corrode very rapidly indeed, leading to a horrible vibration and extreme degradation in efficiency.

One obvious way around this would be to connect the battery the other way around so that it is pumping electrons into the engine block and the propeller and thus saving the latter from corrosion, which indeed it would do - for a while. The problem with this arrangement is that it puts an extreme strain on the battery which is now pumping electrons flat out into what is effectively a 'bottomless pit' and the battery's life will be severely shortened. What, in fact, is done in a properly sorted arrangement, is to have a completely self-contained earth circuit which is not connected in any way to the engine block or hull (if it is a metal one).

...and so to cars

However, this is complex and expensive and while a similar, though less extreme, situation exists in a car, few car manufacturers have felt the need to go to the trouble of fitting a separate earth when they have the convenience of the car's frame and body to form a readily conducting earth circuit. (Those manufacturers who do fit a separate earth are usually specialist builders using glass-fibre bodies where the possibility of using the body as an earth is not handed to them on a plate.)

Before the 1960s when cars were designed by engineers, as opposed to production accountants, the thickness and quality of steel was generally sufficient to allow the use of a positive earth. The resulting corrosion was not significant and battery life was promoted. However, with the constant pressure to economise on materials, manufacturers started to use thinner and thinner guages of steel. Then came a godsend (to them) in the shape of the alternator. With its very much greater output for a unit of similar size and speed when compared with the traditional dynamo, it allowed them to switch over to negative earth, use even thinner steel and rely on the alternator to save the battery.

By the way, this is not merely of academic interest and you should know the polarity of your family car, especially if it was built in the transitionary era of the early- to mid-sixties. If, for example, you fit a more modern car radio designed for use with a negative earth, to a car with a positive earth, you will most probably destroy the radio by blowing all the diodes and transistors in it. It is not enough to quote the model of car to an accessory supplier as many models had their polarity changed in the middle of their production runs.

!

Why does aluminium foil keep things hot (or cold) ?

Heat is transmitted in three ways –

Conduction

A hotter substance in contact with a cooler one will normally transfer some of its heat to the cooler substance. If you put your hand in a bowl of cold water your hand will get colder and the water, albeit imperceptibly, will get warmer.

Convection

Air around a warm object will heat up by the above process of conduction. The warmer air is less dense and so will rise (see *Why does a hot air balloon rise?*). Cooler air will take its place from below and will come into contact with the warm object, so that the process continues until the temperature of the object cools to that of the surrounding air. (The so-called 'radiators' of a central heating system actually heat by convection and, unless you have sprayed them with matt black paint (see below), it would be more logical to call them 'convectors'.)

Radiation

While most of the heat from a traditional coal fire goes straight up the chimney in the form of convection, the warming glow experienced in the room is mainly transmitted by radiation. (Thermal radiation does not need an intervening medium, which means that objects in a vacuum will eventually cool and can transmit heat to other objects even although no conduction nor convection can possibly take place. If this were not the case the Sun would not heat the Earth.)

Shiny surfaces and light colours are poor radiators and absorbers of heat compared with dull surfaces and dark colours. The application of this can be seen in many everyday objects. For example, the header tank at the top of a car's radiator (mainly 'convector', actually) is normally painted matt black to encourage some heat loss through radiation. The enthusiastic car buff who lovingly polishes this back to the underlying brass - or worse still, chromium plates it - is more poseur than engineer unless he has good reason to believe that his engine is over-cooled.

(It is worth pointing out that the same applies to all forms of radiation - not just heat. For example, the surfaces of certain military aircraft which are designed to be 'invisible' to radar are covered in a matt black velour-like material which absorbs the microwaves from the radar transmitter and hence leaves no tell-tale reflection.)

The same principle applies to the way aluminium foil can be used to keep food warm , or to enable it to be heated or cooked, more quickly. If the shiny side is on the inside, it will reflect the radiated heat back at the food which will help to keep it warm, or reduce the cooking time.

Another example of the same phenomenon is a space blanket, used by rescue services in cold climates. This is a wind proof, water proof fabric with a reflective aluminium coating. If the reflective side is wrapped round the casualty, the maximum amount of body heat is retained for as long as possible by reflecting it inwards.

Similarly, if you look inside a vacuum flask you will see that the inside surfaces are silvered. The vacuum cuts out the transmission of heat by conduction and convection but the silvering is there to reduce any transmission by radiation.

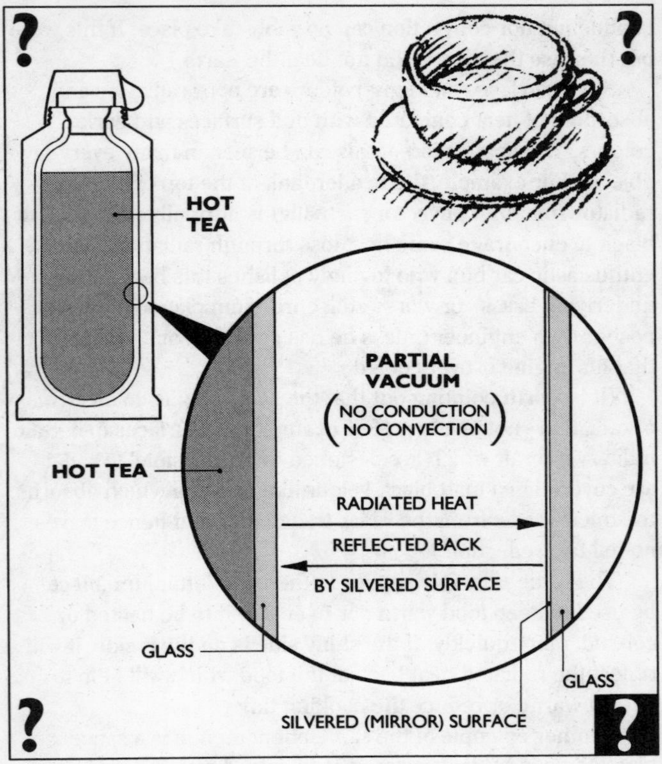

HOT TEA

HOT TEA

PARTIAL VACUUM
(NO CONDUCTION NO CONVECTION)

RADIATED HEAT

REFLECTED BACK

BY SILVERED SURFACE

GLASS

GLASS

SILVERED (MIRROR) SURFACE

...and cold?

However, the properties of aluminium foil are not restricted to keeping things warm. It can also keep things cool.

Shiny side out, wrapped around a plate of sandwiches, the foil will keep them cool by reflecting heat away from the sandwiches and keep them fresher than they would be if they were exposed to the air.

Why does milk turn sour quickly in a thunderstorm ?

Everyone knows that warm weather conditions can lead to milk turning sour if it is left outside a refrigerator. The process is due to micro-organisms called bacteria lactis which are present in milk. A rise in atmospheric temperature increases the growth rate of these bacteria, which then produce lactic acid from lactose, the sugar occurring in mammalian milk. The presence of too much lactic acid causes the protein casein and the lime salts in the milk to separate – the process we call curdling – and the milk turns sour.

The process usually occurs over a period of several hours, but in thundery conditions it is completed much more quickly. The reason – or part of it – is that the life cycle of a storm lasts only one or two hours, and begins when a parcel of air is warmer than the air surrounding it. The warm air then begins to rise and the embryo thunder cell forms as an unstable, warm updraught. As condensation begins to form, latent heat is released so that the whole mass becomes warmer still, until it reaches the point where it is completely out of thermal equilibrium with the surrounding air.

As the cell develops, a cumulus cloud begins to grow, its base extending to a width of about four miles (6.43 km) and its vertical extent to 30 000 feet (9150 m) or more. Within the cloud, strong positive and negative charges of static electricity build up as precipitation particles – water droplets, hail and snow, all moving vigorously up and down on power currents – collide and break up. This continuous disruption of particles builds up enormous electrical charges within the cloud; these continue to grow in size until the difference in electrical potential reaches a critical value depending on the conductivity of the air and the distance between the charges. When the field strength within the cloud approaches one million volts per metre (3.28 ft) a massive electrical discharge takes place – the lighting flash. (See *Why does a good lightning conductor almost never conduct lightning?*)

Usually this discharge takes place from one part of the cloud to another, but sometimes it occurs between cloud and earth. And this is where the other factor that induces milk to turn sour comes in.

As the thundercloud tracks its way across the ground, it carries with it a negative charge of ionized particles. Its path is followed, on the ground, by a locally-induced positive charge that gradually builds up as it progresses. This positive charge climbs trees, steeples, TV aerials or any other object that will bring it closer to the ground as it moves along. When the two charges are close enough, a thin column of electrons reaches down from the cloud and more columns stretch up from the ground to meet them. When the two make contact, an electrical discharge occurs.

As the positive charge moves along the ground, the ions – electrically charged atoms – tend to agitate other atoms in their path, causing friction that generates a release of energy in the form of heat. The rise in temperature is small and brief – but quite enough to accelerate the process that turns milk sour.

Why does ice form on the top of ponds ?

Unlike most substances, which generally contract as they get colder, water behaves most unusually just before it crystallizes into ice. It has its maximum density (of 0.999 973 g/cc) at just below 4 degrees Celsius. As it cools below this temperature it actually expands and at zero degrees Celsius in the liquid state has a density of 0.999 841 g/cc. On freezing, it expands even further and its density drops to 0.916 8 g/cc.

This 'anomalous expansion' of water can be irritatingly inconvenient and is the reason for pipes bursting in winter, tops being forced off milk bottles on frosty mornings, and water-cooled engine blocks cracking when you haven't got round to checking the concentration of the anti-freeze in the autumn.

However, it does mean that, unlike almost all other substances, water's solid form (ice), being less dense, floats on the surface of the liquid form. This is convenient if you enjoy skating although it did have its downside for the Captain of the 'Titanic'.

It is interesting (although probably foolish) to speculate about the history of life on this planet if water were not to behave in this way. Water is in any case one of life's essentials, but given that when the first single-celled organisms evolved, there was no ozone layer nor even significant quantities of atmospheric oxygen, water was a particularly convenient place in which to shelter from the deadly doses of short-wave radiation at the earth's surface. The depths of marshes and ponds must have seemed very stable and attractive environments indeed. To this day many organisms can survive a winter under water even when the surface is covered with ice. If water were to behave like almost any other substance when it solidified, it would sink to the bottom of ponds and effectively kill them as habitable environments.

Why is there almost always a breeze at the seaside ?

Sitting on a beach in the sunshine is often made more comfortable by the fact that there is usually a slight breeze taking the edge off a hot sun. This can, indeed, have its own dangers in that it can hide the strength of the sun's rays and lead to nasty cases of sunburn.

The cause of this breeze is, in fact, the sun itself which heats up the land more quickly than the surface of the sea. Both are covered by air and since hot air rises (See *Why does a hot air balloon rise?*) the hotter air over the land rises and is replaced by the cooler, denser air from above the sea. The result is an onshore sea breeze.

WARM
AIR
COOLING

HEATED AIR OVER
LAND RISING

COOL AIR OVER WATER

SEA BREEZE

However, in the late afternoon the situation reverses. As the sun goes down, the land cools more quickly than the sea which retains any heat that it has gained for much longer. After a lull, there will be an offshore breeze.

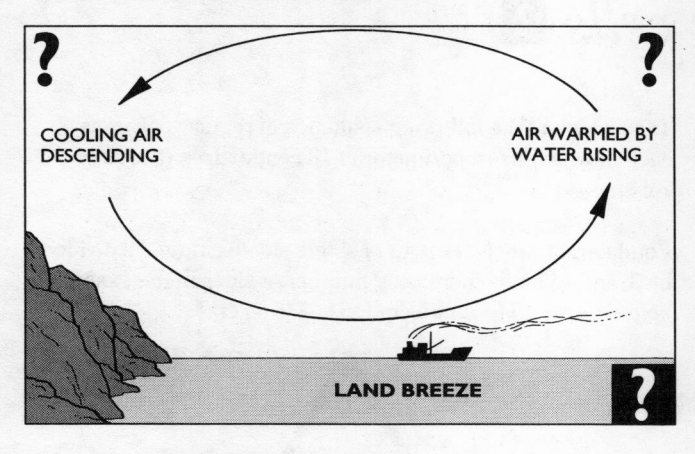

COOLING AIR DESCENDING

AIR WARMED BY WATER RISING

LAND BREEZE

Wind surfers beware!

If this is your sport, it is worth bearing in mind that it is precisely when you are most tired, after a long energetic day, that the wind will tend to drive you away from the safety of the shore.

Why are honeycombs made up of hexagonal cells ❓

Take a look at the following sequence of regular polygons, each of which has a perimeter of 18 centimetres (about 7 inches)–

Equilateral triangle: Length of single side 6 cm (ie. 18 divided by 3, and so on by increasing number of sides in the examples below); Area 15.6 square cm [15.58836 sq cm]

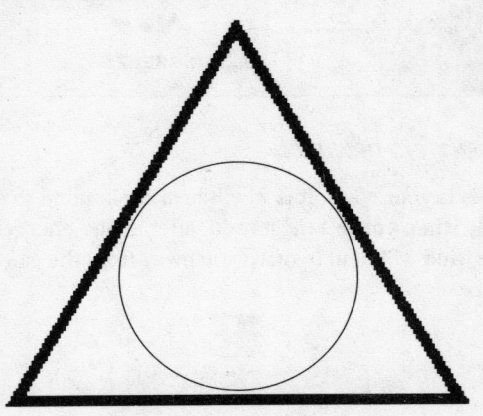

Square: Length of single side 4.5 cm; Area 20.25 square cm

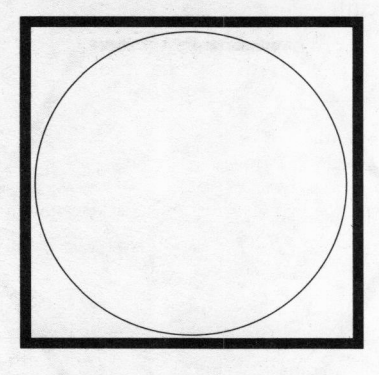

Pentagon: Length of single side 3.6 cm; Area 22.3 square cm
[22.2974208 sq cm]

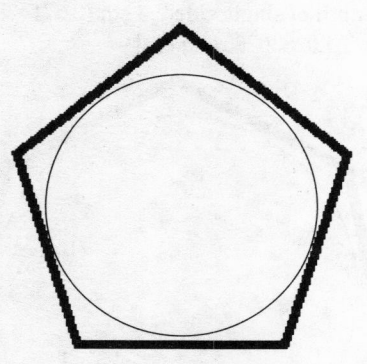

Hexagon: Length of single side 3 cm; Area 23.4 square cm [23.38272 sq cm]

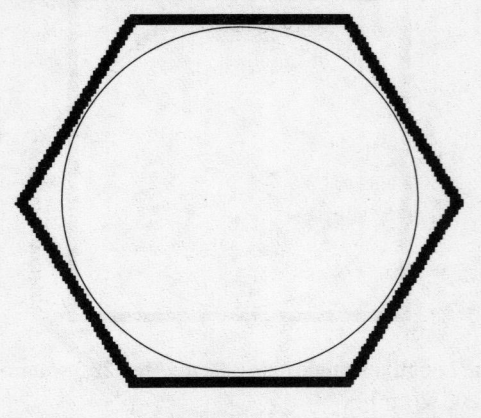

Heptagon: Length of single side 2.6 cm [2.5714285714]; Area 24 square cm [24.028302857 sq cm]

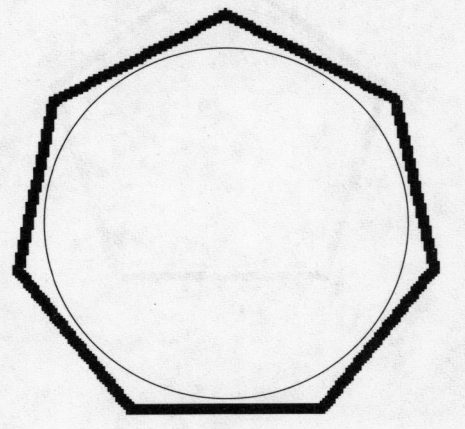

Octagon: Length of single side 2.25 cm; Area 24.4 square cm
[24.443926875 sq cm]

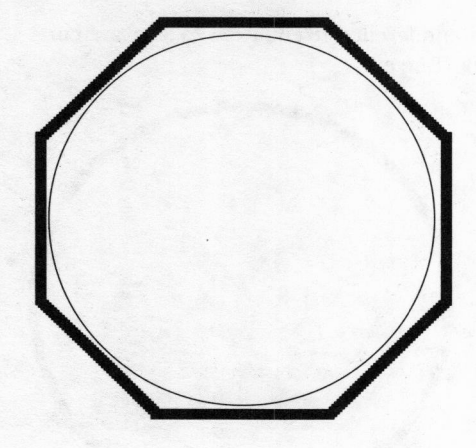

... *and so on.*

Now, consider a circle with an identical perimeter, or circumference, of 18 centimetres -

Circle: Circumference 18 cm; Area 25.8 square cm [25.78310078 sq cm]

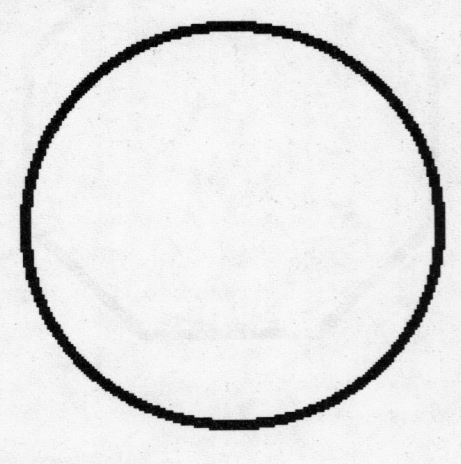

If you are beginning to conclude that, as the number of sides increases and the length of each side decreases, the perimeter becomes a 'more efficient' container of area, you would be absolutely correct. The figures become increasingly 'smoothed' to the ideal shape of a circle, which, by definition, contains the maximum possible area for a given length of line. The adjacent graph helps to illustrate this –

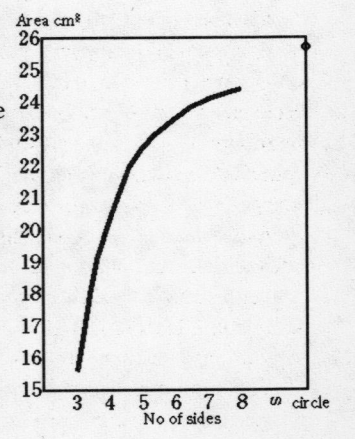

In terms of the most economical use of materials, then, a cell with a circular cross-section would be the most efficient – if it were on its own. Look what happens, however, when you start to pack a lot of circular cells together –

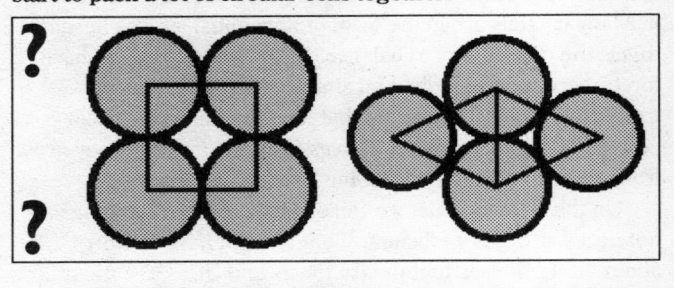

As you can see, if the circles are packed together with the centres in a square arrangement, there is a huge amount of wasted space in the interstices between them. A more efficient – indeed the most efficient, as has been mathematically proved from first principles – arrangement is to pack the centres in the shape of triangles.

However, even in the triangular arrangement (or lattice), there is still wasted space between the circles. Since, in the living world, waste of energy or materials is normally a quick, one-way ticket to extinction, it is not surprising that the bee has found a better way. You will see that in the triangular arrangement, each circle has six points on its surface where it touches one of its neighbours. Imagine pressure being applied to the lattice equally from all directions. The circles would collapse into perfect hexagons as the wasted space is squeezed out.

As there is no possible regular polygon with more than six sides which can pack together without wasted spaces in between it and its neighbours, it stands to reason that a honeycomb is the most efficient possible use of living space and building materials in which to house a hive of bees.

This pattern crops up elsewhere in nature. For example, certain species of lake-dwelling fish which are fiercely territorial but which suffer from overcrowding, tend to form a lattice of hexagonal territories with their neighbours on the bed of the lake. Each fish is maximising the area of its own territory within the constraints of a population unable to take up much room and again this is the most efficient way of doing it. Many crystals are in the form of hexagonal lattices because this is the natural way to balance the attractive and repulsive forces between the individual atoms which are behaving exactly like the fish. The fact that all snow flakes are six-pointed stars, for instance, derives from the fact that they grow from a tiny hexagonal crystalline 'seed'.

On the human scale, we can see that modern buildings are not necessarily very efficient. If one did not have to worry about fitting houses together in towns and cities, the most economical use of materials would be to build towers with a circular ground plan which would include the greatest space for the fewest number of building blocks - and incidentally lose the least heat through the outside wall. (The Picts probably worked this out when building their defensive brochs as they needed very thick walls made up of stones painstakingly piled-up by hand.) A square or rectangular ground plan is about the least efficient in terms of maximising area. Indeed, only the triangle is worse. The most rational ground plan is, in fact, the hexagon with a lattice of hexagonal rooms within it. (As many people who have lived in converted lighthouses or windmills would appreciate, cupboards do not have to be adapted to the curvature of the walls in a hexagonal room as, whatever its size, they will fit in the corners so long as the backs of the cupboards have an angle of 120 degrees.)

So why hasn't it been done? In fact, it has. As early as the 1920s the American engineer, R. Buckminster Fuller, who was appalled at the wastage of building materials in the United States, proposed the idea of the hexagonal house and indeed several have been built to his plans.

Why do people seem to have red eyes in some photographs taken using a flashgun ❓

Family groups worthy of the special effects department of Hammer Films are probably familiar to most amateur photographers. What causes this phenomenon and what can be done to avoid it?

Light behaves rather in the way a billiard ball bounces off the cushion of a billiard table or the way the ball bounces off the bat in an electronic game such as 'Breakout'. If it hits an object straight on at a right angle, it will bounce straight back. If it strikes an object at any other angle, it will be reflected at that angle but in the opposite direction.

The bright light from a flashgun, if it hits the eye straight on, is reflected off the very fine blood vessels at the back of the eye and is bounced straight back through the lens of the camera producing the 'redeye' effect which ruins so many group photographs.

To get round this, it is desirable to use a 'bounce flash' which angles the light upwards and bounces it down off the ceiling of the room. This, in any case, gives a softer, more diffused and natural light, which avoids the rather bleaching effects of direct flash. However, if you don't have a bounce flash, you can generally avoid the redeye effect by introducing even the smallest angle between the camera and the subject - and by telling the subject not to look at the camera. The reason for this is that as well as being reflected, light is also refracted or 'bent' when it travels at an angle from a less dense medium (the air) to a more dense one (the fluid within the eyeball).

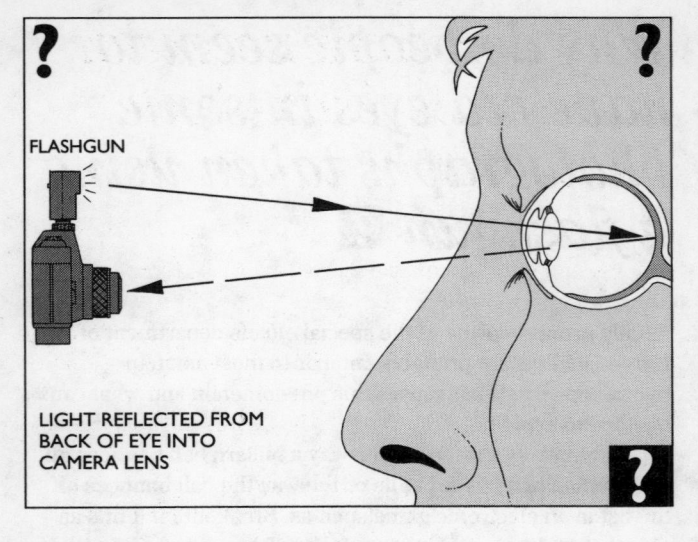

FLASHGUN

LIGHT REFLECTED FROM BACK OF EYE INTO CAMERA LENS

The light rays are also refracted again - but in the opposite direction - when they are reflected out of the eyeball. Given the net total of these effects you will be extremely unlucky indeed to suffer from the redeye effect in your photographs at anything other than a 'direct hit'.i.e.the light is bouncing *straight* back.

Why are there no green mammals (nor black plants) ❓

This question began to intrigue us when we read press reports of an international conference held in London in the summer of 1991 which was specifically set up to debate this issue. The conference came up with no firm conclusions and we certainly aren't offering any here. This entry is by far the most speculative one in the whole of this book, but that doesn't stop us having one or two thoughts on the matter.

In a world where there are so many striking examples of natural camouflage, it does, at first sight, seem odd that no mammals have evolved which could exploit the competitive advantage of blending into the predominantly green background of (say) a tropical rain forest. Insects, reptiles, amphibians and birds are past masters at this art. (In fact, certain species of sloth do achieve a sort of green coloration, but this is effected through the benefits of green algae which live in their fur). However, it remains true, to the best of anyone's knowledge, that there are no naturally-pigmented green mammals. Indeed, the concept is so strange to the human mind that it has become almost a metaphor of that which is alien - the proverbial little green men of space fantasy.

However, perhaps we are wrong to find this seeming lack of evolutionary adaptation surprising.

Mammals are relatively late arrivals on the scale of evolutionary time. The beginnings of life on this planet date from about 3.5 billion years ago, with the earliest animals dating from about 800 million years back. The earliest mammals came on the scene only about 65 million years ago. Even taking the several bouts of extinction into account, many species have had an evolutionary 'head start' on mammals, especially when you also consider the relative slowness of breeding of most mammals.

Coupled with this fact, it seems worth remembering just how rare mammals are. Of the over one million, major, living (ie. those not lost through extinction) animal species known to science only about 4000 of them are mammals. (Compare this with over 290,000 known species of beetle.)

Perhaps because we are mammals ourselves and because, until recently, the majority of our domesticated animals (with a few exceptions such as the hen and the silk-worm) are also mammals, we tend to over-rate the mammalian significance in the scheme of things.

Perhaps our small number on the face of the Earth simply has not given us the time nor the statistical 'clout' to produce some green examples.

And black plants?

Perhaps mammals need not feel inferior about this, however. Green plants (and planktons) depend upon light for the production of their energy through the process known as photosynthesis. Green plants appear green because they absorb all parts of the spectrum with the exception of green, which, of course, they reflect. Since the amount of light that a plant can absorb is often a limiting factor on its growth, it would seem more efficient if they were to absorb light of all colours and consequently appear black.

A superficial, but unsatisfactory, answer is that chlorophyll, the pigment necessary for photosynthesis is, in itself, green. This only begs the question as to why chlorophyll isn't black...

(We are grateful to Richard Thwaites of Norwich, England, for raising the latter issue in the Letters page of the *New Scientist*. The debate has continued in letters to the *New Scientist* with one suggestion that if plants were black, they would absorb too much radiation and effectively 'cook' themselves. See *Why should a kettle be shiny or white?*)

Epilogue

Fool: ...The reason why the seven stars are no more than seven is a pretty reason.

Lear: Because they are not eight?

Fool: Yes, indeed. Thou wouldst make a good fool.

William Shakespeare
King Lear, I,v,38